ᴛʜᴇ**new** biology

Stem Cell Research

Revised Edition

ᵀᴴᴱ**new** biology

Stem Cell Research

Medical Applications and Ethical Controversies
Revised Edition

JOSEPH PANNO, PH.D.

☑Checkmark Books®
An imprint of Infobase Publishing

STEM CELL RESEARCH: Medical Applications and Ethical Controversies, Revised Edition

Copyright © 2010, 2005 by Joseph Panno, Ph.D.

Checkmark Books
An imprint of Infobase Publishing
132 West 31st Street
New York NY 10001

.

Library of Congress Cataloging-in-Publication Data
Panno, Joseph.
    Stem cell research : medical applications and ethical controversies / Joseph Panno.—Rev. ed.
        p. cm. — (The new biology)
    Includes bibliographical references and index.
    ISBN 978-0-8160-6851-7 (hardcover)
    ISBN 978-0-8160-8330-5 (pbk)
    1. Stem cells—Research. 2. Stem cells—Research—Moral and ethical aspects. I. Title.
    QH588.S83P36 2011
    616'.02774—dc22—2009030506

Checkmark Books are available at special discounts when purchased in bulk quantities for businesses, associations, institutions, or sales promotions. Please call our Special Sales Department in New York at (212) 967-8800 or (800) 322-8755.

You can find Facts On File on the World Wide Web at http://www.factsonfile.com

Text design by Erik Lindstrom
Illustrations by the Author
Photo research by Diane K. French
Composition by Hermitage Publishing
Cover printed by Bang Printing, Brainerd, Minn.
Book printed and bound by Bang Printing, Brainerd, Minn.
Date printed: July 2010
Printed in the United States of America

10 9 8 7 6 5 4 3 2 1

This book is printed on acid-free paper.

# Contents

Preface

When the first edition of this set was being written, the new biology was just beginning to come into its potential and to experience some of its first failures. Dolly the sheep was alive and well and had just celebrated her fifth birthday. Stem cell researchers, working 12-hour days, were giddy with the prospect of curing every disease known to humankind, but were frustrated by inconsistent results and the limited availability of human embryonic stem cells. Gene therapists, still reeling from the disastrous Gelsinger trial of 1998, were busy trying to figure out what had gone wrong and how to improve the safety of a procedure that many believed would revolutionize medical science. And cancer researchers, while experiencing many successes, hit their own speed bump when a major survey showed only modest improvements in the prognosis for all of the deadliest cancers.

During the 1970s, when the new biology was born, recombinant technology served to reenergize the sagging discipline that biology had become. This same level of excitement reappeared in the 1990s with the emergence of gene therapy, the cloning of Dolly the sheep, and the successful cultivation of stem cells. Recently, great excitement has come with the completion of the human genome project and the genome sequencing of more than 100 animal and plant species. Careful analysis of these genomes has spawned a new branch of biological research known as comparative genomics. The information that scientists can now extract from animal genomes is expected to improve all other branches of biological science. Not to be outdone, stem cell researchers have found a way to produce embryo-like stem cells from ordinary skin cells. This achievement not only marks the end of the great stem cell debate, but it also provides an immensely powerful procedure, known as cellular dedifferentiation, for studying and manipulating the very essence of a cell. This procedure will become a crucial weapon in the fight against cancer and many other diseases.

The new biology, like our expanding universe, has been growing and spreading at an astonishing rate. The amount of information that is now available on these topics is of astronomical proportions. Thus, the problem of deciding what to leave out has become as difficult as the decision of what to include. The guiding principle in writing this set has always been to provide a thorough overview of the topics without overwhelming the reader with a mountain of facts and figures. To be sure, this set contains many facts and figures, but these have been carefully chosen to illustrate only the essential principles.

This edition, in keeping with the expansion of the biological disciplines, has grown to accommodate new material and new areas of research. Four new books have been added that focus on areas of biological research that are reaping the benefits of genome science and modern research technologies. Thus, the New Biology set now consists of the following 10 volumes:

1. *Aging, Revised Edition*
2. *Animal Cloning, Revised Edition*
3. *Cancer, Revised Edition*
4. *The Cell, Revised Edition*
5. *Gene Therapy, Revised Edition*
6. *Stem Cell Research, Revised Edition*
7. *Genome Research*
8. *The Immune System*
9. *Modern Medicine*
10. *Viruses*

Many new chapters have been added to each of the original six volumes, and the remaining chapters have been extensively revised and updated. The number of figures and photos in each book has increased significantly, and all are now rendered in full color. The new volumes, following the same format as the originals, greatly expand the scope of the New Biology set and serve to emphasize the fact that these technologies are not just about finding cures for diseases but are helping scientists understand a wide range of biological processes. Even a partial list of these revelations is impressive: detailed information on every gene and every protein that is needed to build a human being; eventual identification of all cancer genes, stem cell–specific genes, and longevity genes; mapping of safe chromosomal insertion sites for gene therapy; and the identification of genes that control the growth of the human brain, the development of speech, and the maintenance of mental stability. In a stunning achievement, genome researchers have been able to trace the exact route our human ancestors used to emigrate from Africa nearly

65,000 years ago and even to estimate the number of individuals who made up the original group.

In addition to the accelerating pace of discovery, the new biology has made great strides in resolving past mistakes and failures. The Gelsinger trial was a dismal failure that killed a young man in the prime of his life, but gene therapy trials in the next 10 years will be astonishing, both for their success and for their safety. For the past 50 years, cancer researchers have been caught in a desperate struggle as they tried to control the growth and spread of deadly tumors, but many scientists are now confident that cancer will be eliminated by 2020. Viruses, such as HIV or the flu, are resourceful and often deadly adversaries, but genome researchers are about to put the fight on more rational grounds as detailed information is obtained about viral genes, viral life cycles, and viruses' uncanny ability to evade or cripple the human immune system.

These struggles and more are covered in this edition of the New Biology set. I hope the discourse will serve to illustrate both the power of science and the near superhuman effort that has gone into the creation and validation of these technologies.

Acknowledgments

I would first like to thank the legions of science graduate students and postdoctoral fellows who have made the new biology a practical reality. They are the unsung heroes of this discipline. The clarity and accuracy of the initial manuscript for this book was much improved by reviews and comments from Diana Dowsley, Michael Panno, Rebecca Lapres, and later by Frank K. Darmstadt, executive editor, and the rest of the Facts On File staff. I am also indebted to Diane K. French and Elizabeth Oakes for their help in securing photographs for the New Biology set. Finally, as always, I would like to thank my wife and daughter for keeping the ship on an even keel.

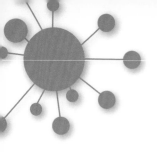

Introduction

Stem cells are special cells that have the ability to divide for an indefinite period and can give rise to a wide variety of specialized cell types. This ability, known as developmental plasticity, is a common feature of fertilized eggs and early embryonic cells (known as blastomeres). A fertilized egg, being able to give rise to all of the body's cells, has the highest degree of developmental plasticity and, thus, is said to be totipotent. Blastomeres are also totipotent, but this level of plasticity decreases quickly. Consequently, blastomeres from a five-day-old human embryo (consisting of about 200 cells) can only give rise to a limited range of cell types and, accordingly, are said to be pluripotent. As development progresses, individual cells become multipotent (able to give rise to only a few cell types) before assuming their final form as a specialized cell that can only give rise to other cells of its kind. Stem cells may be isolated from embryos, umbilical cords, and adult tissues. When they originate from embryos or umbilical cords, they are equivalent to pluripotent blastomeres. Stem cells isolated from adult tissues are usually multipotent, but pluripotent forms have been identified. Most recently, stem cells have been produced by reprogramming skin cells, an advance that is expected to revolutionize the field.

When placed in culture, stem cells grow and divide indefinitely, and scientists are learning how to coax them into producing cell types that may be used to cure many diseases. The extraordinary powers and versatility of these cells have generated an interest level that approaches a fever pitch. Stem cell therapies may be able to treat cardiovascular disease, spinal cord disorders, Parkinson's disease, Alzheimer's disease, and some cancers. Leukemia, a cancer affecting white blood cells (WBCs), is already being treated by replacing the cancerous cells with stem cells programmed to differentiate (transform) into healthy WBCs. Diseases that affect the brain, spinal cord, or heart are ideal candidates for stem cell therapy because these organs have lost the talents that stem cells retain; in particular, the ability to proliferate (grow and reproduce) and differentiate.

All eukaryote cells (organisms containing nuclei and organelles) at some point in their lives possess the powers of reproduction and differentiation, but those powers become a liability when cells are trying to live as a community. This is particularly true for a community as complex as an animal's body. The human brain, for example, is an intricate assemblage of 100 billion neurons that is constructed during embryonic development; once established, this network would be destroyed if the cells continued dividing. The neurons in our brain can form new associations with other neurons throughout our life, but they become post-mitotic (lose the ability to divide) soon after an individual is born. Many other organs, such as the spinal cord, heart, kidneys, and muscles, adhere to the same developmental pattern: active cell division during embryogenesis, loss of cell division in the adult.

The loss of cell division in an animal's body is a trade-off that allows the cell community to produce organs of a predictable size and shape. But if a person suffers a disease or trauma such as a heart attack, the post-mitotic cells, in this case myocytes (cardiac muscle cells), are unable to repair the damage. If the damage is extensive, the heart muscle cannot contract properly and the patient dies or has to have a transplant. However, some of our tissues and organs, such as skin, liver, and bone marrow, retain the power of division throughout the life span of the individual. If a finger is cut, the wound is able to heal because cells in the skin divide to fill the gap. If liver cells die, or a portion of the organ is removed surgically, the cells will divide and grow to repair the damage. Red blood cells (RBCs), with a life span of only 120 days, have to be replaced on a daily basis. In this case, it is stem cells, located in the bone marrow where blood cells are made, that divide and differentiate into both red and white blood cells, thus replacing them as they wear out.

Scientists believe that if stem cells can replace worn-out red blood cells it might be possible to train them to repair organs, such as the brain or heart, that are incapable of repairing themselves. If scientists are successful, we may live to see the day when there is no more cardiovascular disease, no more brains wasting away in an Alzheimer's fog, and quadriplegics will rise out of their chairs and walk again. However, many investigators believe that these results can only be obtained by using embryonic stem cells. But harvesting these cells requires the destruction of human embryos, a practice that many believe is immoral and unethical. Consequently, stem cell research has become a very contentious area of research that has roused an often-rancorous debate at all levels of society. Recent success with stem cells isolated from adult tissues and stem cells created from skin cells is expected to calm this debate.

Stem Cell Research, Revised Edition, one title in the New Biology set, discusses the different types of stem cells, how they are studied in the laboratory, and the diseases that may be treated with them.

This new edition, now with color photographs and line drawings, has been extensively revised and expanded.

The first four chapters are new and take a different approach to stem cell research from the first edition. Chapter 1 discusses the origin and evolution of ordinary cells and their remarkable ability to form plants and animals. This chapter places stem cells within the much broader context of cell communication and embryological development. Chapter 2 provides a detailed discussion of human and animal stem cells. Much attention is paid to the differences and similarities between embryonic and adult stem cells. Therapeutic cloning, a controversial hybrid of embryonic stem cell therapy and animal cloning, is the focus of chapter 3. The ongoing ethical debate and a sensational case of research fraud involving this therapy have cast serious doubt on the future of this technology. Chapter 4 is focused on a new form of stem cell that is produced by reprogramming ordinary skin cells. These cells, first described in 2007, could make embryonic stem cell research and therapeutic cloning obsolete.

Many of the medical applications of stem cell therapy, discussed in chapter 5, were still on the drawing board when the first edition of this book was published, but are now routine medical therapies or are being tested in hundreds of clinical trials in the United States and Europe. The cost of developing these therapies increases every year, and consequently the role of pharmaceutical companies is more important than ever. Three of these companies are profiled in chapter 6.

Stem cell therapies have great potential, but scientists frequently run well ahead of the practical considerations. Stem cells, particularly those from embryos, may be able to treat many diseases, but they can also be extremely dangerous and unpredictable. These concerns are discussed at length in chapter 7. Although the ethical issues of stem cell research, as discussed in chapter 8, have changed little over the years, the legal status is an ongoing debate. As discussed in chapter 9, some European laws regarding therapeutic cloning and embryonic stem cells have changed dramatically over the years, and many believe similar changes are about to occur in the United States as well.

The final chapter, as before, provides background material on cell biology, biotechnology, and other topics that are relevant to stem cell research. The cell biology and biotechnology primers presented in this chapter have been extensively revised and condensed in the hope that the reader will be able to obtain the necessary background information as quickly as possible.

Cells through Time

Cells are often described as being the basic units of life. Although true, the word *unit* leaves an unfortunate impression that cells are no more than a mechanical component in a much larger machine and that the true attributes of life are part of the grander structure rather than the individual cells. But this is far from true. Cells are very much alive and are capable of surprisingly complex behaviors—they grow and reproduce, they hunt and forage, they build plants, animals, and other incredibly complex organisms, and they possess an elaborate system for communicating with the outside world and with other cells. Their ability to communicate enables the hunters to distinguish prey from kin and was essential for the formation of the first cell colonies, from which plants and animals evolved. For more than 1 billion years, single cells dominated Earth's biosphere, and even today, on a weight basis, they still do. Understanding a cell, and especially a stem cell, is best done from an evolutionary perspective. The extraordinary nature of a cell is a direct consequence of the many challenges it had to face throughout its long history.

THE ORIGIN OF LIFE

Life on Earth arose from microscopic bubbles that formed spontaneously in the oceans about 3.5 billion years ago. How these bubbles formed and how they evolved into living organisms are the subject of much research and debate. There are of course no fossils from that period that could be used to reconstruct the events that led to the appearance of the first cell, so scientists rely on what is known about the ancient Earth and about cells, alive today, that have retained ancient characteristics.

The young Earth was hot, with surface temperatures high enough to melt lead. The atmosphere consisted of methane, ammonia, and carbon dioxide, with barely a trace of oxygen. These gases were released into the atmosphere by volcanic eruptions and outgassing from the rocks. After 500 million years, Earth's surface temperature dropped considerably, stimulating a pelting rain that lasted thousands of years, covering most of the planet in water just as it is now. Heat and the electrical activity of the storms converted the atmospheric gases into molecules such as sugar, amino acids, nucleic acids, and fatty acids. The heat fused many of these molecules into macromolecules (chains of molecules) such as protein, deoxyribonucleic acid (DNA), ribonucleic acid (RNA), and a fatty substance called phospholipid. These compounds enriched the water, turning the oceans into a nutrient broth. The phospholipid could not dissolve in the water the way the other macromolecules could but spread out on the surface, producing Earth's first oil slick.

Storms whipped up the surface of the young oceans into a riot of waves that broke on the primeval shores. The waves generated billions of microscopic bubbles, some of which were stabilized by a thin phospholipid coating or membrane. When the bubbles formed,

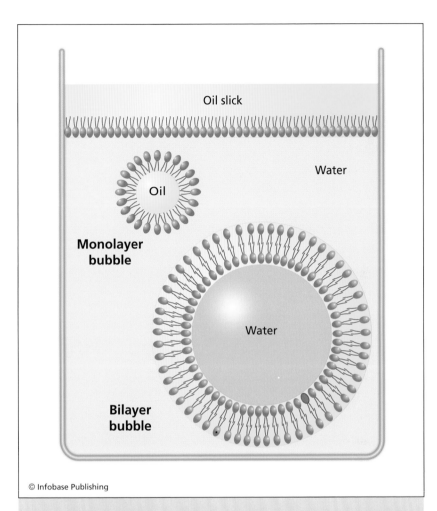

© Infobase Publishing

Phospholipid bubbles. Phospholipid molecules have a hydrophilic head-end (red ovals) and two hydrophobic tails that do not mix with water and will avoid being surrounded by it. In an oil slick, the hydrophobic tails mix with the oil while the heads stay close to the water. In turbulence, phospholipids form two kinds of bubbles: a monolayer that can only capture a drop of oil and a bilayer that can capture a drop of water. The bilayer allows the hydrophobic tails to associate with themselves, while the heads associate with water on both the inside and outside surfaces of the bubble.

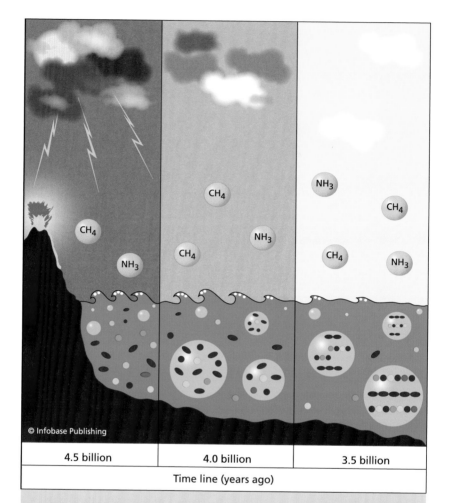

| 4.5 billion | 4.0 billion | 3.5 billion |

Time line (years ago)

Origin of the first cells. Organic molecules essential for life were synthesized spontaneously 4.5 billion years ago when the Earth was hot, stormy, and wracked with constant volcanic eruptions. Some of the organic molecules were captured by lipid bubbles (light blue spheres) formed by ocean turbulence near a shoreline, and by 3.5 billion years ago the first cells learned how to assemble the molecules into a variety of polymers. Nucleic acids, amino acids, fats, and sugars were among the organic molecules produced in the prebiotic oceans; only the nucleic acids (colored circles) and amino acids (brown ovals) are shown. Major gases in the atmosphere included methane (CH_4) and ammonia (NH_3).

they each captured a tiny drop of the nutrient broth that was all around and, for a short time, were like a billion separate laboratories all experimenting with a unique combination of the materials at hand. When the bubbles burst, they released the fruits of their labor into the water for the next generation to capture. After countless generations, the bubbles reached a point where they could regulate their internal environment and could perform a few simple chemical reactions that stabilized their membranes and organized their internal functions. Eventually they learned how to store information in a primitive genome and to control their own duplication, rather than depending on ocean waves and turbulence. After 500 million years (1 billion years after the Earth was formed), simple bubbles evolved into a population of cells, the first life-form to appear on Earth.

THE FIRST CELLS

The first cells are believed to have had an RNA genome that contained a dozen or so genes. These genes coded for very simple proteins that may have been used for structural purposes or as enzymes that could extract energy from the available nutrients. The cells simply absorbed the ready-made molecules from the water in what is known as a heterotrophic lifestyle. Initially, there was plenty of food to go around. Scientists have estimated that the young biosphere had enough dissolved nutrients to last a few million years. This is a long time, but not nearly long enough to account for the emergence of higher organisms, an event that took 2 billion years.

How did the early cell populations continue on after the original pool of nutrients had been consumed? Scientists believe the biosphere was rescued from certain doom by the evolution of the first cells into more advanced forms known as the ancestral prokaryotes, which consisted of both heterotrophs and autotrophs. Unlike the heterotrophs, autotrophs get their energy directly from the Sun in a process called photosynthesis. This transition from heterotrophy to autotrophy was critical to the survival of life on Earth. Dying and

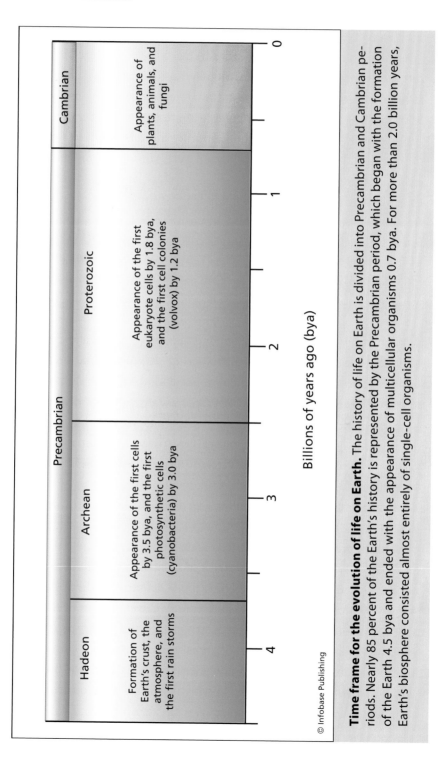

Billions of years ago (bya)

Time frame for the evolution of life on Earth. The history of life on Earth is divided into Precambrian and Cambrian periods. Nearly 85 percent of the Earth's history is represented by the Precambrian period, which began with the formation of the Earth 4.5 bya and ended with the appearance of multicellular organisms 0.7 bya. For more than 2.0 billion years, Earth's biosphere consisted almost entirely of single-cell organisms.

decomposing autotrophs replenished the dissolved nutrients that the heterotrophs needed to survive. The heterotrophs, functioning like scavengers, kept the system healthy by minimizing the buildup of decomposed material. Once this simple ecosystem was established, the ancestral prokaryotes evolved into archaea, bacteria, and eukaryotes, the three major divisions of life in the world. The eukaryotes went on to form protozoans, fungi, plants, and animals. Stem cell research is concerned almost entirely with this type of cell.

Bacteria and archaea are prokaryotes that have a relatively simple structure but a complex biochemistry (prokaryote means "before the nucleus"). Typically, these cells have a genome consisting of about 4,000 genes. The archaea, as the name suggests, are the most ancient form of cellular life known to exist on Earth today. These prokaryotes can live in very extreme environments, common to the young Earth, such as volcanic vents where there is little or no oxygen and the temperature may exceed 158°F (70°C). Bacteria, being more advanced than the archaea, are better adapted to environments provided by Earth today; they are comfortable in the presence of oxygen and generally prefer cooler temperatures ranging from 68°F (20°C) to 104°F (40°C). Bacteria were also the first autotrophs to appear on Earth. These bacteria usually occur as chains and are known as Cyanobacteria or blue-green algae.

Eukaryotes are bigger and much more complex than prokaryotes. These cells are strictly aerobic (require oxygen), generally prefer mild temperatures, and have developed the autotrophic lifestyle to an exquisite degree. The cellular compartment, homogeneous in prokaryotes, is divided into the nucleus and the cytoplasm (eukaryote literally means "true nucleus"). The nucleus contains the DNA, which may encode as many as 30,000 genes. The cytoplasm is further divided into a complex set of organelles, which are responsible for most of the cell's biosynthetic activity. All eukaryotes have special cell-surface structures that made the appearance of multicellular organisms possible. The cytoplasm of an autotrophic eukaryote,

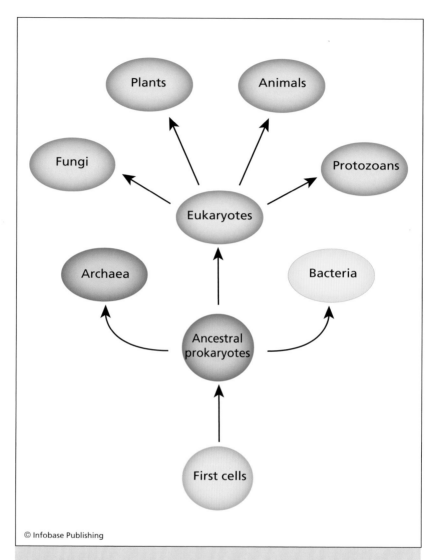

Cell classification. The first cells evolved into the ancestral prokary-otes, which gave rise to the archaea, bacteria, and eukaryotes, the three major divisions of life in the world. The archaea and bacteria are very similar anatomically, but differ biochemically. Eukaryotes, anatomically and biochemically distinct from both the archaea and bacteria, gave rise to plants, animals, protozoans, and fungi.

such as *Chlamydomonas*, is dominated by the green chloroplast, an organelle that is responsible for the cell's photosynthetic activity. Additional information regarding the structure and function of the eukaryotes is provided in chapter 10.

The evolution of the eukaryotes was as important to the survival of the young biosphere as was the appearance of the first autotrophs. Prokaryote heterotrophs served a valuable role as scavengers, but they could not limit the size of the autotroph population, and hence the young biosphere was unstable. Single-cell eukaryotes, known as protozoans, stabilized the system by developing the ability to hunt autotrophs as a source of food, thus marking the beginning of predator-prey relationships and the stability that such relationships bring to an ecosystem. Protozoans developed special cell-surface structures that were used to hunt and capture prey. These structures not only made it possible for them to communicate with their environment and with other cells, but they paved the way for cell colonies, plants, and animals.

HOW CELLS COMMUNICATE

An elaborate structure called the glycocalyx contains all of the cell's communication hardware. The glycocalyx is a forest of glycoproteins and glycolipids that covers the surface of every cell like trees on the surface of the Earth. This structure can be found among the prokaryotes, but the eukaryote glycocalyx is much more complex. Scientists have estimated that a typical eukaryote has a glycocalyx consisting of more than 1,000 different kinds of glycoproteins. The earliest eukaryotes hunted bacteria for food, and their single-cell descendants, the protozoans, still make a living in this way. Scientists believe the glycocalyx evolved over time to meet the demands of this kind of lifestyle. Viewed in this way, the glycocalyx is indeed versatile, providing hardware for communications and for the capture and ingestion of food molecules and cells. In addition, some of the glycoproteins are specially adapted for holding cells together

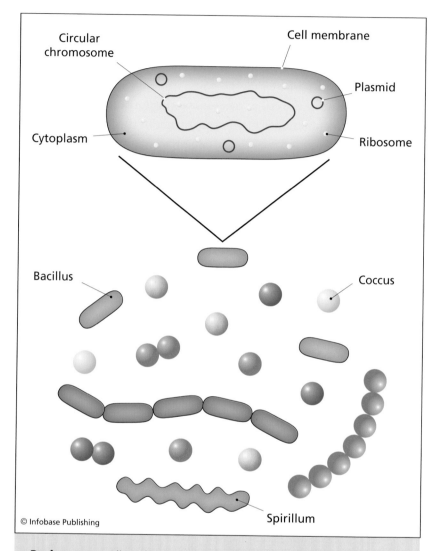

Circular chromosome

Cell membrane

Plasmid

Cytoplasm

Ribosome

Bacillus

Coccus

Spirillum

© Infobase Publishing

Prokaryotes. All prokaryotes have the same basic anatomy, consisting of a cell membrane, cytoplasm (or protoplasm), and a circular DNA chromosome. Some bacteria have a second, smaller chromosome called a plasmid, which may be present in multiple copies. The cytoplasm contains a wide assortment of enzymes and molecules as well as ribosomes (protein-RNA complexes that are involved in protein synthesis). The cells may be spherical (coccus), rod-shaped (bacillus), or wavy corkscrews (spirillum), appearing singly, in pairs, or linked together into chains.

and played an important role in the evolution of multicellular organisms.

The communication hardware, built primarily from the glycoproteins, consists of signaling pathways that relay information from the membrane to the interior of the cell. There are many different kinds of pathways, each designed to detect a different external signal. Bacteria, for example, consume a variety of sugars and have different pathways for detecting glucose, lactose, and sucrose. Binding of these food molecules to a receptor activates the signaling pathway, which leads to the formation of a cytoplasmic second messenger. The second messenger may activate enzymes in the cytoplasm, or it might move into the nucleus where it stimulates the expression of one or more genes. The activated cytoplasmic factors and/or the gene products are then directly responsible for initiating the ingestion and utilization of the food.

Protozoan hunters evolved cell-surface receptors designed to detect and bind preferred species of bacteria. This is possible because the bacterial glycocalyx, which is species-specific, provides the recognition factors for the hunters. Thus, the ability of one cell to detect the presence of another cell is very ancient, preceding the emergence of multicellular organisms. This ability is crucial for the survival of all animals because it forms the basis for the immune system, which protects an individual from invading parasites, viruses, and a host of microbes.

In animals, communication between cells is mediated by paracrine (direct cell-to-cell communication) and endocrine (indirect) systems. Both of these systems rely on the release of molecules from one cell or group of cells, which modify the behavior of other cells. In a paracrine system, cells release signaling molecules into the immediate neighborhood, thus affecting a small, localized population of cells. In an endocrine system, signaling molecules, known as hormones, are released into the blood, where they are able to affect the behavior of cells and tissues throughout the body. White blood

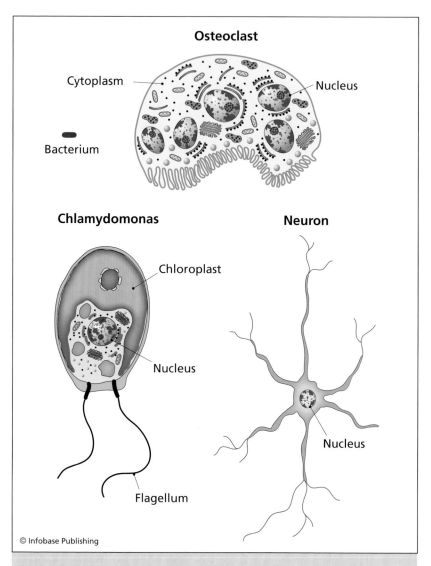

Osteoclast

Cytoplasm

Nucleus

Bacterium

Chlamydomonas

Neuron

Chloroplast

Nucleus

Nucleus

Flagellum

© Infobase Publishing

Eukaryotes. The prokaryotes gave rise to large and very complex cells called eukaryotes. A single bacterium (upper left) is shown for scale. Osteoclasts are multinucleated cells involved in bone remodeling in humans and other animal species. Neurons are found in all animals, where they form the brain, spinal cord, and nerve fibers. Chlamydomonas is a photosynthetic protozoan, a large group of eukaryotes that gave rise to all multicellular plants. Non-photosynthetic (heterotrophic) protozoans gave rise to animals.

Paramecia. This is a type of freshwater protozoan, characterized by an ovular shape, cilia, and an oral groove for feeding. (*Gary Retherford/Photo Researchers, Inc.*)

cells of the human immune system coordinate their attack on invading microbes by signaling one another with molecules known as interleukins. This is an example of a paracrine cell signaling system because the interleukins are released into the local environment and are not released into the circulatory system. On the other hand, the pituitary gland, central to the vertebrate endocrine system, releases hormones into the blood, affecting cells and tissues throughout the body. This kind of communication is intended for the coordination of an animal's physiology.

Paracrine and endocrine signaling are efficient and effective, but higher organisms would never have evolved if there were no other way for cells to communicate with one another. The problem with these systems is that they are relatively slow. If the human brain had to use the endocrine system in order to wiggle a toe, it would take several seconds before the order could be received and acted

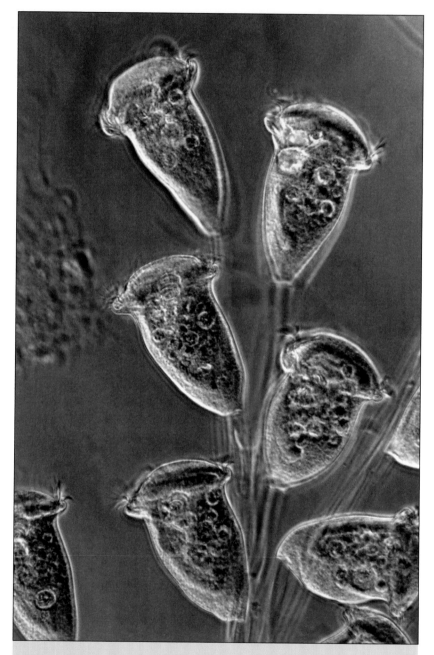

Sessile ciliates. These are protozoans living in water and moist soil. (*Eye of Science/Photo Researchers, Inc.*)

Volvox colonies. Light micrograph of colonies of *Volvox aureus,* under phase contrast illumination. Volvox has been variously classified as a green alga or as a flagellate protozoan. Each colony is a large hollow sphere made up of thousands of cells. The cells are arranged in a single layer at the surface of the sphere and are connected by fine cytoplasmic threads. Each cell has two flagella, which point outward and enable the colony to swim in its freshwater environment. Daughter colonies (dense green areas) are produced asexually by splitting within the colony. Magnification: ×80 when printed 10cm high. (*Sinclair Stammers/Photo Researchers, Inc.*)

upon. Fortunately, there is a third way for cells to communicate, which depends on the coordination of ion channels: In this system, a signaling molecule or an electrical stimulus opens ion channels, which allows the inward movement of positively charged sodium ions. The inward movement of these charged particles stimulates

(continues on page 19)

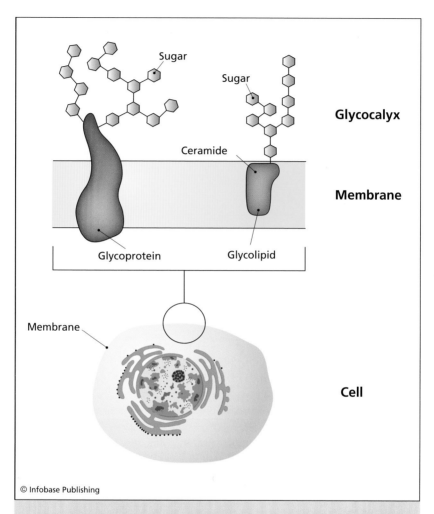

Sugar

Sugar

Glycocalyx

Ceramide

Membrane

Glycoprotein

Glycolipid

Membrane

Cell

© Infobase Publishing

The eukaryote glycocalyx. The eukaryote's molecular forest consists of glycoproteins and glycolipids. Two examples are shown at the top, a glycoprotein on the left, and a glycolipid on the right. The glycoprotein trees have "trunks" made of protein and "leaves" made of sugar molecules. Glycolipids also have "leaves" made of sugar molecules, but the "trunks" are a fatty compound called ceramide that is completely submerged within the plane of the membrane. The glycocalyx has many jobs including cell-to-cell communication and the transport and detection of food molecules. It also provides recognition markers so the immune system can detect foreign cells.

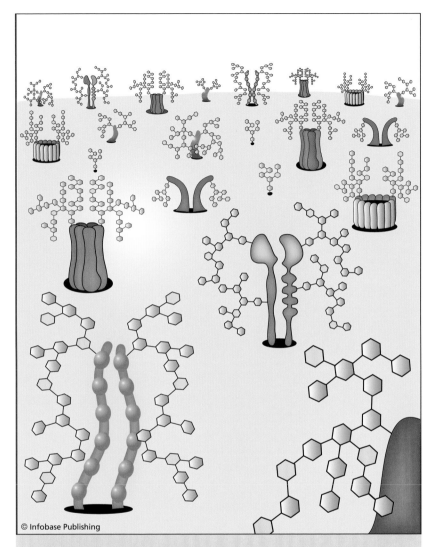

A panoramic view of the glycocalyx. This image is a three-dimensional rendering of the glycoproteins and glycolipids that cover the surface of all cells. The paired glycoproteins in the foreground are cadherin (front left) and integrin (upper right). These structures help hold cells together in animal tissues. The barrel-like structures are part of the cell's communication hardware.

© Infobase Publishing

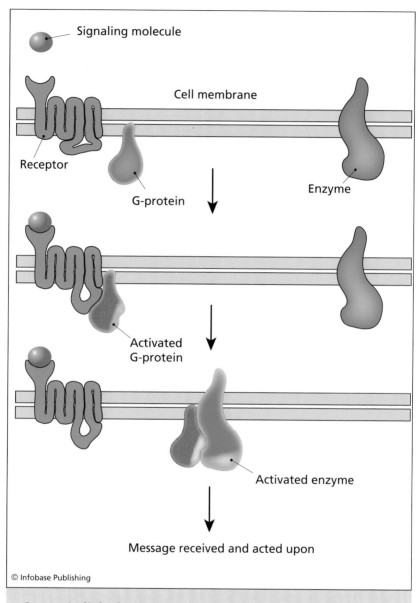

© Infobase Publishing

G-protein-linked receptor. The signaling molecule binds to the receptor, leading to the activation of the G-protein, which in turn activates an enzyme. The enzyme, in turn, activates secondary messengers, which deliver the signal to the appropriate destination.

(continued from page 15)

the cell to release a special kind of paracrine molecule, known as a neurotransmitter. The neurotransmitter binds to other neurons, which activates their ion channels and, subsequently, the release of more neurotransmitters. In this way, the initial signal can pass through an entire circuit in a fraction of a second. Using this system, the human brain can order the big toe to move in less than a millisecond. This form of cell communication was invented by neurons and forms the basis for all neural communications involving the brain, spinal cord, peripheral nerves, and neuromuscular junctions.

THE RISE OF MULTICELLULAR ORGANISMS

The evolution of the glycocalyx made it possible for early eukaryotes to hunt for food. But it also made it possible for those cells to communicate with one another for the purpose of forming small colonies. Initially, the colonies consisted of about a dozen cells that would associate with one another for brief periods in order to accomplish a specific task, either to enhance their hunting ability or for reproduction.

It took almost 2 billion years for the first colonies to evolve into true multicellular organisms. The first of these were similar to modern-day red algae, a very simple form of seaweed. The red algae were followed by the marine sponges, which consist of different kinds of cells but lack tissues and a nervous system. The next stage of development came with the cnidarians (jellyfish and corals), animals that are made from only eight cell types but that possess a mouth, digestive tract, and tissues such as epithelia and muscles. Echinoderms (sea stars and sea cucumbers) and mollusks (snails, clams, and octopods) introduced many refinements including a central nervous system, a visual system, and enhanced locomotion. All of these innovations were refined with the appearance of the vertebrates during the Cambrian period about 530 million years ago.

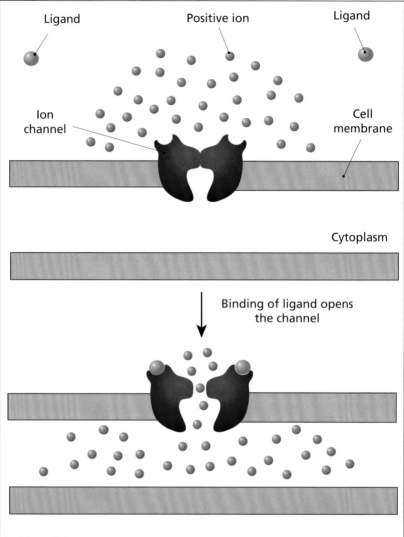

Ligand

Positive ion

Ligand

Ion channel

Cell membrane

Cytoplasm

Binding of ligand opens the channel

© Infobase Publishing

Ligand-gated (Lg) ion channel. The closed channel keeps positively charged ions outside the cell, and in this state, the membrane is said to be polarized (i.e., positive on the outside, relative to the inside). The channel opens when bound to a signaling molecule (ligand), allowing the ions to rush inside, thus depolarizing the membrane and initiating an electrical current.

The evolutionary transition from single cells to multicellular organisms is replayed every time a fertilized egg develops into a multicellular plant or animal. This process, known as embryogenesis, produces the many different kinds of cells that form a plant or an animal. These cells are then molded into the overall body plan of the organism, containing a variety of tissues and organs. A notable characteristic of the body's cells is that they all possess the same genes, the same genome, despite their radically different shapes, sizes, and functions. This seemingly paradoxical characteristic will be explained in the next chapter.

THE MAMMALIAN EMBRYO

Mammalian embryos, like all vertebrate embryos, pass through similar developmental stages that are marked by the appearance of the morula, blastula (or blastocyst), gastrula, and neurula. There are, of course, many differences between vertebrate species with respect to embryogenesis. Mammals develop inside the mother's womb, where they receive nourishment through the placenta and are surrounded by a chorionic membrane that produces a fluid-filled incubation chamber. The embryos of nearly all other vertebrates, including fish, amphibians, reptiles, and birds, develop inside eggs, which contain all the nourishment they need. Consequently, all of the cells in a frog or bird embryo become part of the adult.

This is not the case with mammalian embryos, since a portion of their cells must be used to produce the placenta and the chorion. The distinction between embryonic and non-embryonic cells is evident by the blastula stage, when two kinds of cells have developed: the inner cell mass (ICM) and the trophoblast. Embryonic tissue is derived exclusively from the ICM, while the trophoblast differentiates into the placenta and the chorionic membrane. The trophoblast also performs an essential role in the implantation of the embryo in the mother's uterus. The cells of the ICM are called blastomeres, but

(continues on page 25)

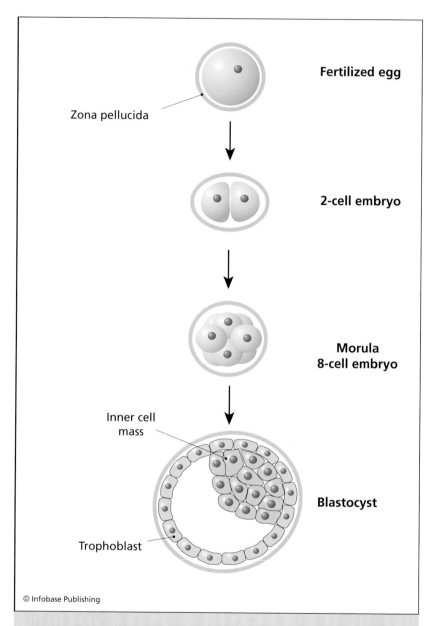

Zona pellucida

Fertilized egg

2-cell embryo

Morula
8-cell embryo

Inner cell
mass

Blastocyst

Trophoblast

© Infobase Publishing

Mammalian embryogenesis up to the blastula stage. Mammalian embryos, unlike amphibians, are surrounded by non-embryonic cells that make up the zona pellucida and trophoblast, the latter giving rise to the placenta and chorionic membrane. The embryonic cells are in the inner cell mass (ICM). Note that mammalian blastulas are also called blastocysts.

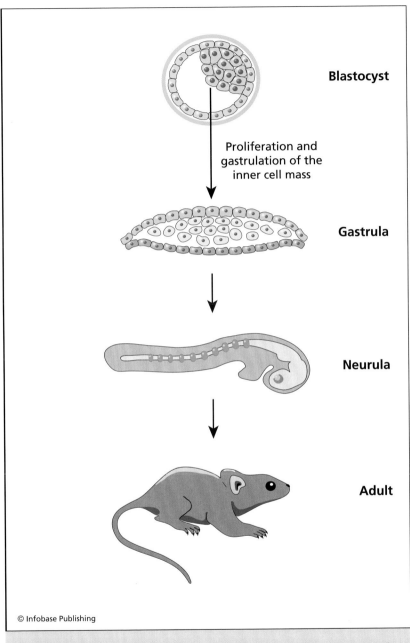

Blastocyst

Proliferation and
gastrulation of the
inner cell mass

Gastrula

Neurula

Adult

© Infobase Publishing

Later stages of mammalian embryogenesis. After the morula
stage, the embryo develops into a blastula (or blastocyst), gastrula,
and neurula.

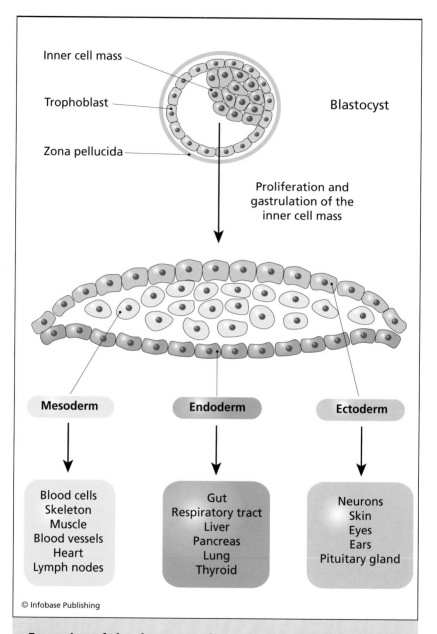

Inner cell mass

Trophoblast

Zona pellucida

Blastocyst

Proliferation and gastrulation of the inner cell mass

Mesoderm

Endoderm

Ectoderm

Blood cells
Skeleton
Muscle
Blood vessels
Heart
Lymph nodes

Gut
Respiratory tract
Liver
Pancreas
Lung
Thyroid

Neurons
Skin
Eyes
Ears
Pituitary gland

© Infobase Publishing

Formation of the three germ layers in a mammalian embryo. During gastrulation, movement and proliferation of the inner cell mass (ICM) produces the three primary germ layers: the mesoderm, endoderm, and ectoderm. Eventually, the ectoderm covers the entire embryo. Movements and fate of the trophoblast and zona pellucida are not shown.

(continued from page 21)
when isolated and grown on a culture plate they are called embry-onic stem cells.

The mammalian blastula is called a blastocyst because it em-beds itself in the lining of the uterus. The mammalian gastrula has a flattened profile, in contrast to the spherical shape of amphibian and invertebrate embryos. Flattened embryos are a common feature among animals, such as birds and reptiles, that produce very yolky eggs, on top of which the embryo floats like a tiny raft. Mammals do not develop inside an egg but, being descended from reptilian stock, have retained their form of embryogenesis. The earliest signs of neu-rulation involve the formation of an anatomical feature known as the primitive streak. This feature, which appears in humans when the embryo is 14 days old, has been used by legislators as a decision boundary. That is, once the primitive streak appears, the embryo cannot be used as a source of stem cells.

An important property of embryonic development is cell-to-cell contact and communication. This intimate and very intricate relationship between the cells is crucial for the induction and coor-dination of cellular differentiation. By touching one another, or by signaling at a distance, cells can induce other cells to differentiate into a particular type of tissue. By careful regulation of the timing and location of induction, the cells are able to establish the embryo's basic body plan: where is up, where is down, which end is going to be the head, and which will be the tail.

Embryonic induction occurs in three major stages, called pri-mary, secondary, and tertiary induction. Primary induction is as-sociated with gastrulation, a coordinated movement of the cells that leads to the formation of the three germ layers: the ectoderm, meso-derm, and endoderm, which give rise to all the tissues of the adult body. Secondary induction involves a complex interaction between the three germ layers to initiate neurulation, which involves the development of the brain, spinal cord, segmented spinal column, and peripheral nerves. Tertiary induction regulates organogenesis, or development of the body's organs and appendages.

Stem Cells

Drs. Ernest McCulloch and James Till, then professors at the University of Toronto and the Ontario Cancer Institute, discovered stem cells in the 1960s. These cells are unique, having the ability to differentiate into more than one kind of cell, a characteristic called plasticity. All stem cells, regardless of their source, have three general properties: They are capable of dividing and renewing themselves for long periods; they are unspecialized; and they can give rise to specialized cell types.

The highest level of plasticity is called totipotency, meaning the cell can differentiate into all the cells of a given organism. Fertilized egg cells (oocytes) are the only cells possessing this degree of plasticity. Stem cells invariably possess lesser degrees of plasticity, being either pluripotent (can produce most of the organism's cells) or multipotent (can differentiate into only a few cell types). The plasticity of a stem cell varies depending on whether it originates from an embryo or from an adult organism. In general, stem cells from embryos are pluripotent, whereas stem cells from adults are multipotent. Embryonic stem cells have been isolated from mice, monkeys, and humans, but most of what is known about these cells has been learned by studying mouse stem cells.

Throughout the 1990s, when scientists were studying stem cells from rodents, standard protocols were developed for culturing, testing, and manipulating these cells. Other stem cells, from different species or from adult tissues, are now studied by using those protocols so that one type of stem cell can be easily compared with another. The protocols cover the behavior of stem cells in vivo (within a living organism) and in vitro (in cell culture). The most important in vitro characteristic includes the cell's ability to proliferate (grow and divide) for an indefinite period of time, while maintaining an embryonic phenotype. Phenotype, used in this context, refers to all the observable characteristics of the cell: its shape, or morphology (in this case, a simple, rounded shape); its behavior, meaning the way in which it interacts with other cells; and, finally, the composition of the glycocalyx, the molecular forest that covers the surface of all cells. The types of proteins embedded in the membrane of an embryonic cell are different from those of a fully differentiated adult cell. Thus, the composition of the glycocalyx varies depending on the differentiation state of the cell. To gain an appreciation for the magnitude of this change, imagine an aerial view of a coniferous forest compared to an aerial view of a deciduous forest.

An important in vitro characteristic of stem cells is their ability to differentiate into many different kinds of cells. Differentiation may occur spontaneously or through a process called directed differentiation, which occurs when the cells are allowed to contact one another or when certain growth factors are added to the culture medium. The in vivo behavior of candidate stem cells is established by isolating the cells and then injecting them into a mouse to see if they will differentiate. How the stem cells respond to these

STEM CELL NOMENCLATURE

CELL	ABBREVIATION	ALTERNATE
adult stem cell	AS cell	ASC
bone marrow stem cell	BMS cell	—
cardiac stem cell	CS cell	CSC
embryonic stem cell	ES cell	ESC
embryonic germ cell	EG cell	EGC
endothelial progenitor cell	EP cell	EPC
induced pluripotent stem cell	iPS cell	iPSC
umbilical cord stem cell	UCS cell	—
hematopoietic stem cell	HS cell	HSC
human embryonic stem cell	HES cell	hESC
mesenchymal stem cell	MS cell	MSC

Note: Some of the stem cells shown in the table are discussed in other chapters. Observe the following: 1) The European spelling of "hematopoietic" is "haematopoietic"; 2) hemopoietic is synonymous with hematopoietic; 3) a bone marrow stem cell is equivalent to a hematopoietic stem cell; 4) endothelial progenitor cells and mesenchymal stem cells are subsets of hematopoietic stem cells found in the bone marrow.

procedures depends primarily on whether they were isolated from an embryo or an adult.

Studying human embryonic stem cells is a difficult and demanding enterprise, not only because of the science, but because of the greater social issues associated with any experiment in which a human being, no matter at what stage of development, is the guinea pig. Ethical issues in research of this kind set the pace and the scope of the experiments that are allowable. At the present time, stem cell research is focused on embryonic stem (ES) cells, adult stem (AS) cells, therapeutic cloning, and induced pluripotent stem (iPS) cells. These iPS cells are the most recent addition to the stem cell galaxy and are expected to revolutionize the field. This chapter will focus on ES and AS cells. Therapeutic cloning and iPS cells will be the subject of later chapters.

CELLS PRODUCED BY STEM CELL DIFFERENTIATION

CELL	DESCRIPTION
adipocyte	cells that make and store fat compounds
astrocyte	a type of glia (glue) cell that provides structural and metabolic support to the neurons
cardiomyocyte	cells that form the heart, also called myocytes
chondrocyte	cells that make cartilage
dendritic cell	antigen-presenting cell of the immune system
endothelial cell	cells that form the inner lining (endothelium) of all blood vessels
hematopoietic cell	cells that differentiate into red and white blood cells
keratinocyte	cells that form hair and nails
mast cell	associated with connective tissue and blood vessels
neurons	cells that form the brain, spinal cord, and peripheral nervous system
oligodendrocyte	myelin-forming glia cells of the central nervous system
osteoblast	give rise to osteocytes, or bone-forming cells
pancreatic islet cells	endocrine cells that synthesize insulin
smooth muscle	muscle that lines blood vessels and the digestive tract

EMBRYONIC STEM CELLS

The ultimate stem cell is the fertilized egg that being totipotent can give rise to an entire organism consisting of hundreds of different kinds of cells. Human, mouse, and amphibian blastomeres, from

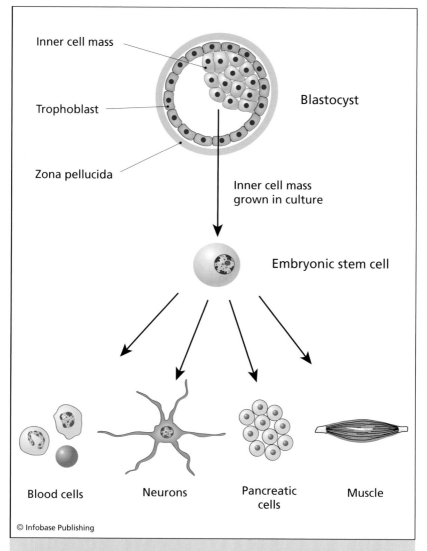

Differentiation of embryonic stem cells. Embryonic stem cells are obtained from the inner cell mass of a blastocyst. When cultured, these cells can differentiate into many different kinds of cells, representing the three germ layers.

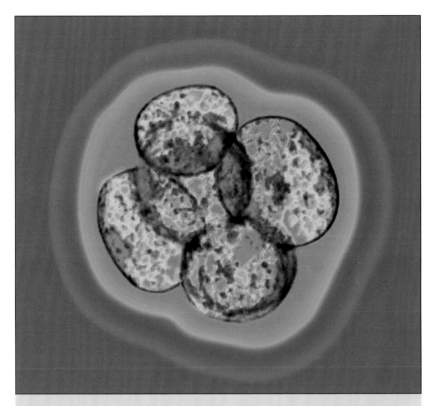

Micrograph of a human embryo soon after fertilization. The cells, or blastomeres, result from divisions of the fertilized egg and are surrounded by the protective zona pellucida layer. The cells of embryos like these will continue to divide before implanting into the wall of the uterus. There they will continue to grow to form a human fetus composed of millions of cells. (Andrew Paul Leonard/Photo Researchers, Inc.)

two- or four-cell embryos, also retain their totipotency and are good examples of embryonic stem cells.

Mammalian ES cells are obtained exclusively from the inner cell mass (ICM) of a blastocyst, and when placed in cell culture they can differentiate into many kinds of cells, representing all three embryonic germ layers. However, once the association between the ICM and the trophoblast is disrupted (as when ES cells are placed in culture), the ES cells cannot develop into an embryo. For this reason, they are said to be pluripotent rather than totipotent. In culture, ES cells can proliferate for years while retaining an embryonic phenotype.

Most of what is known about stem cells has come from studying mice and rats. All of the protocols investigators now use to evaluate stem cells come from work on those animals, and many scientists continue using rodents to gain a deeper understanding of what it is that makes a stem cell tick. But stem cells from mice cannot be used to cure a human disease. Repairing a severed spinal cord or curing Parkinson's disease can only be attempted with human stem cells. Consequently, many scientists are eager to study human stem cells and are hoping they will show the same properties of plasticity and proliferation demonstrated by rodent stem cells.

Colored scanning electron micrograph of a teratoma formed by human embryonic stem cells. Magnification: ×1,100 when printed 10cm wide. (Prof. Miodrag Stojkovic/Photo Researchers, Inc.)

Scientists first demonstrated the remarkable plasticity of ES cells in the 1980s when cells from the ICM of a mouse blastocyst were transferred to the cavity of a second mouse blastocyst, where they differentiated, in vivo, into a variety of tissues. In other experiments, ES cells were placed in culture dishes and allowed to differentiate spontaneously (in this case, the cells are said to have differentiated in vitro). The first, and crucial, stage of in vitro differentiation involves the aggregation of the cells into small clumps called embryoid bodies. Contact between the cells is necessary for differentiation to occur and echoes the events of normal embryogenesis, in which cell-to-cell contact and interactions between the three germ layers determine the developmental fate of a given group of cells.

In culture, cell-to-cell communication within the embryoid body leads to the formation of neurons, skin cells, contracting muscle tissue, and other cell types. Although embryoid bodies have a loose organization, some of them resemble blastocysts. Something similar happens when ES cells are injected under the skin of adult mice. Again, the cells tend to aggregate into small clumps, which in this case are benign tumors called teratomas. Examination of these tumors has shown that they consist of gutlike structures, neural tissue, cartilage, bone, and sometimes hair.

When cultured ES cells aggregate to form embryoid bodies, they are trying to form a gastrula and the three germ layers, just as they would have done during normal embryonic development. But without the trophoblast surrounding them and the signals they normally receive after implanting in the mother's womb, these cells are like small children trying to find their way home on a very dark night. They have lost their vision and have no map to guide them. They can make all of the cells the body will ever need, but they do not know where to put them or how to connect them.

Collecting Human Embryonic Stem Cells

Dr. James Thomson at the University of Wisconsin collected human embryonic stem cells for the first time in 1998. Thomson's group

Portrait of James Thomson, developmental biologist at the University of Wisconsin-Madison. In 1998, his research group reported the first successful culture of human embryonic stem cells. (*University of Wisconsin-Madison / Photo Researchers, Inc.*)

obtained cells from five-day-old blastocysts that were donated by in vitro fertilization (IVF) clinics with the permission of the parents. The cells obtained from the blastocysts are the blastomeres of the ICM, representing the entire embryo. When collected and grown in cell cultures, they are known as embryonic stem cells.

Thomson's team began with 36 embryos, of which only 14 developed to the blastocyst stage. The ICM was isolated from these embryos and used to establish five human ES cell lines: H1, H7, H9, H13, and H14. (There are now more than 120 hESC lines worldwide.) Culturing of these cells was carried out in a multistep process. After removal of the ICM from the blastocyst, the cells were placed in culture dishes containing a layer of feeder cells. The ES cells were allowed to grow for two weeks, after which they were transferred to plates lacking a feeder layer in order to induce the formation of embryoid bodies. Cells from the edge of the embryoid bodies were collected and transferred to fresh plates.

The five original cell lines continued to divide without differentiating for six months and were able to form teratomas in mice. Cell line H9 went on to proliferate for more than two years and is now being used by many research groups around the world. Karyotyping of the cultures showed that three of the cell lines were male (XY) and two were female (XX). All of the cell lines continued to maintain a normal karyotype. This later observation was extremely

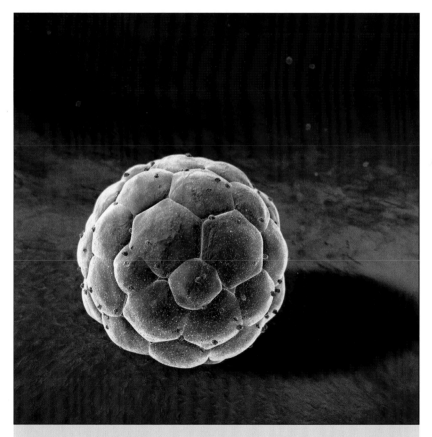

Computer illustration of a blastocyst embryo about to implant on the wall of the uterus. (*hybrid medical animation/Photo Researchers, Inc.*)

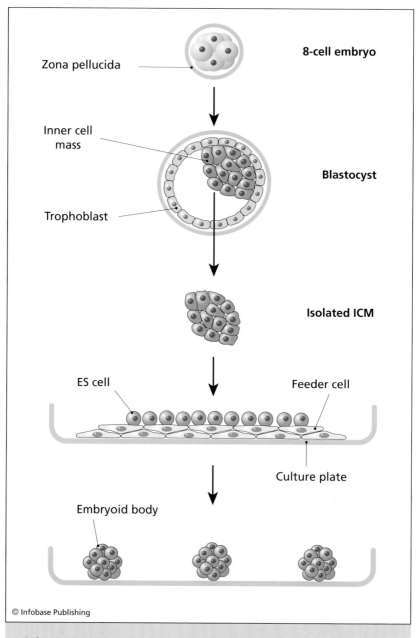

Zona pellucida

8-cell embryo

Inner cell
mass

Blastocyst

Trophoblast

Isolated ICM

ES cell

Feeder cell

Culture plate

Embryoid body

© Infobase Publishing

Culturing embryonic stem cells. The ICM is harvested from a blastocyst, dissociated into single cells, and grown in a culture plate containing mouse feeder cells. To stimulate differentiation and the formation of
embryoid bodies, the cells are transferred to plates lacking feeder cells.

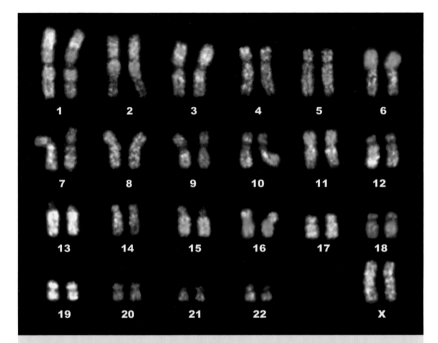

Fluorescence light micrograph of the 46 chromosomes from a normal human female. Each cell contains 22 matched pairs of chromosomes and one sex-determining pair (bottom right). Female and male karyotypes differ only in the sex pair. Male sets would be labeled XY instead of the X pair here. Magnification unknown. (*L. Willatt/Photo Researchers, Inc.*)

important, since if the cells grew well in culture but suffered chromosomal damage (i.e., broken chromosomes or daughter cells that received an incorrect number of chromosomes) they would be useless for medical therapies. Genetic abnormalities are, nevertheless, still a problem in prolonged cultures.

Other researchers have established different lines of stem cell cultures, consisting of embryonic germ (EG) cells, by collecting cells from the gonadal ridge of eight- to 12-week-old fetuses. The cells from each fetus were used to establish a separate culture line. The gonadal ridge is located in the fetuses' lower mid-back. Cells in this area form the gonads of the adult, which produce eggs or

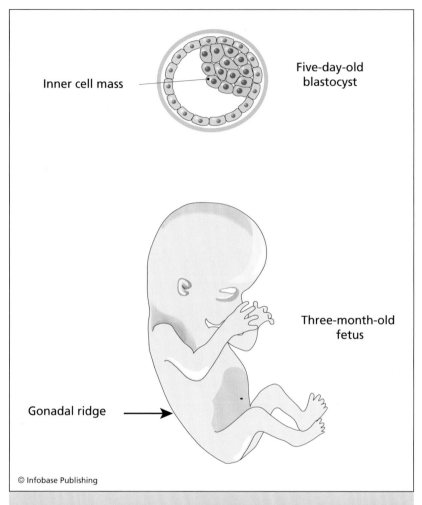

Human embryo and fetus used for the collection of stem cells.
Embryonic stem cells are collected from the inner cell mass (ICM) of
five-day-old blastocysts. Embryonic germ cells are collected from the
gonadal ridge of two- to four-month-old fetuses. At three months of
age, the fetus is about two inches (50 mm) long.

sperm (i.e., germ cells). The culture conditions are usually similar
to those used by Thomson's team, except for the inclusion of a cytokine called leukemia inhibitory factor (LIF) and a mitogen called

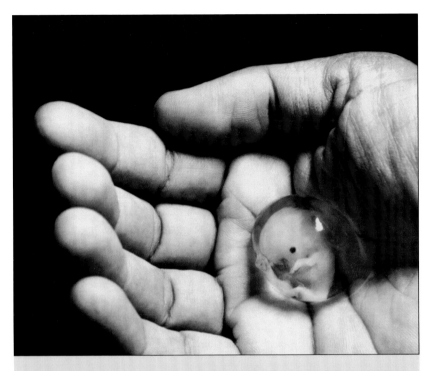

Eight-week human fetus resting in an adult's palm. The complete fetus is visible inside an intact amniotic sac, the membranous, fluid-filled bag that cushions and protects the developing infant. At eight weeks of gestation all of the fetus's organs are present and function-ing. At this stage it is no longer called an embryo. (Dr. M. A. Ansary/ Photo Researchers, Inc.)

fibroblast growth factor (FGF). These compounds are added to the culture plates to stimulate growth and cell division.

After three weeks of growth, the human EG cells resemble mouse ES cells and express several embryonic markers, such as stage-specific embryonic antigen-1 (SSEA-1) and SSEA-3. Both of these markers are cell-surface glycoproteins. Some of the cultures were induced to form embryoid bodies by the removal of LIF, FGF, and the feeder cell layer. Examination of the embryoid bodies showed differentiation of cell types representing all three embryonic germ layers. The identities of the cells were based on physical appearance and the expression of various markers.

The EG cells failed to produce teratomas in mice, suggesting that they do not have the developmental plasticity of ES cells and hence are less desirable candidates for stem cell therapies. In addition, everything that can be achieved with EG cells can be accomplished with embryonic or adult stem cells, calling into question the value of this line of research. Many people are willing to accept the isola-

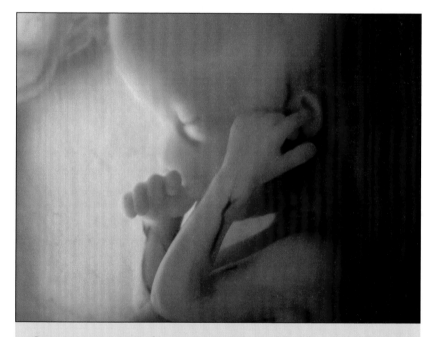

Computer-generated image reconstructed from scanned human data. This image presents a computer-generated left-sided view of a 12-week-old fetus. The age is calculated from the day of fertilization. This image reveals the developmental regions of the head and face. The eyelids have formed and are fused so that the eyes are closed. The hands can be observed with clearly distinctive fingers. (Anatomical Travelogue/Photo Researchers, Inc.)

tion of stem cells from human blastocysts that are only a few days old but are opposed to the use of highly developed human fetuses as a source for these cells, particularly if they can be obtained in some other way.

Fate of the Donor Embryo

Until recently, there was no way to remove cells from the ICM of a human blastocyst without killing the embryo. But in January 2008, researchers at Advanced Cell Technology (ACT) in Worcester, Massachusetts, reported some success with harvesting one or two blastomeres without killing the embryo. It is too soon to tell if this procedure is really an improvement over the original. The problem with this new method is that while the experimental procedure itself does not kill the embryo, it will die nonetheless. The embryos used at ACT were donated by IVF clinics for research purposes (with the permission of the parents), and thus it is extremely unlikely that they will ever be carried to term. The researchers could collect the blastomeres and then return the embryos to the freezer, in the hope that they will someday be carried to term. But given the expense and the inefficiency of IVF, parents will likely choose embryos that have not been tampered with. Consequently, it appears that this new method of harvesting blastomeres has no practical application.

The situation involving the fetuses used to collect EG cells is somewhat different in that they are killed during the abortion procedure. Several methods are available for terminating a pregnancy, and the amount of damage done to the fetal tissue varies depending on which procedure is used. However, the choice of method is up to the attending physician and cannot, by law, be influenced by a researcher who is planning to collect fetal tissue. The use of human fetuses for stem cell research is carefully regulated by the American Public Health Service Act, and scientists must obtain special permission to conduct these experiments, including informed consent from the parents.

ADULT STEM CELLS

Stem cells that are collected from adult tissues or organs are called adult stem (AS) cells. The term also applies to stem cells collected from children, infants, and even umbilical cord blood. Not long ago, scientists believed all repairs in the adult body were carried out by the affected tissue: If the skin is cut, other skin cells along the damaged area divide and migrate to seal the wound; break a leg, and chondrocytes (bone-forming cells) repair the damage. Other organs, such as the brain and heart, were thought to be incapable of self-repair since the myocytes and neurons were known to be post-mitotic. The picture, however, has become more complex and, from a clinical point of view, much more interesting now that AS cells have been identified.

The first adult stem cell to be characterized was found in the bone marrow. These cells are called hematopoietic stem cells (HS cells) and are known to be responsible for replenishing blood cells. The bone marrow contains other types of stem cells, such as endothelial progenitor cells (EPCs) and mesenchymal stem cells (MSCs). The EPCs have the potential to differentiate into endothelial cells (cells that line blood vessels and the digestive tract), and the MSCs can form many types of cells, including bone, fat, and cardiomyocytes (heart cells). It may be that most repairs are carried out by stem cells and that organs such as the brain and heart, with the help of specialized stem cells, may have some ability to heal themselves.

While ES cells are defined and identified by having been isolated from the ICM of a blastocyst, identifying adult stem cells and determining their source are very difficult. Indeed, no one knows the ultimate source of these cells. Some scientists have suggested that they are embryonic cells, set aside during development of each tissue, while others believe they may have been part of a migrating

ADULT TISSUES AND ORGANS KNOWN TO HAVE STEM CELLS

SOURCE	DESCRIPTION
bone marrow	These occur as hematopoietic stem cells, which give rise to red and white blood cells. The bone marrow also contains mesenchymal stem cells, which differentiate into cartilage and bone, and endothelial progenitor cells, which differentiate into blood vessels and cardiomyocytes.
brain	Stem cells of the brain can differentiate into the three kinds of nervous tissue—astrocytes, oligodendrocytes, and neurons—and, in some cases, blood-cell precursors.
digestive system	Located in intestinal crypts, or invaginations. These stem cells are responsible for renewing the epithelial lining of the gut.
heart	Stem cells have been isolated from the heart, where they reside in small clusters. The extent to which these cells repair the heart is unknown.
pancreas	Many types are believed to exist, but examples have yet to be isolated. Some neural stem cells are known to generate pancreatic β cells.
skeletal muscle	These stem cells may be isolated from muscle or bone marrow. They mediate muscle growth and may proliferate in response to injury or exercise.
skin	Stem cells of the skin are associated with the epithelial cells, epidermal cells, hair follicle cells, and the basal layer of the epidermis. These stem cells are involved in repair and replacement of all types of skin cells.

population of embryonic cells that took up residence in various parts of the body during the processes of neurulation and organogenesis. A third possibility is that stem cells were produced after embryonic development was complete by the dedifferentiation of a select group of cells within the various tissues of the body. It is not clear why these cells are able to repair some tissue but not others.

The plasticity of AS cells and ES cells has been demonstrated by determining the fate of both kinds of cell after being injected into mice. ES cells, being undifferentiated, do not exhibit a tendency to find "home"—that is, to return to the tissue from which they derive. Instead, ES cells associate with one another, forming teratomas in various regions of the body. AS cells, on the other hand, have differentiated enough that they know where home is and that is where they collect: Stems cells derived from bone marrow return to the bone marrow and those of neural origin migrate to the brain or spinal cord. In general, researchers have shown that ES cells, grown in culture, can differentiate into a wide variety of cell types, representing all germ layers, whereas cultured AS cells differentiate into a smaller range of cells, representing one or two germ layers.

The limited versatility of AS cells has been challenged by Dr. Catherine Verfaillie in a series of studies she conducted at the University of Minnesota and more recently at the Catholic University of Leuven in Belgium, where she serves as the director of the Stem Cell Institute. In 2002, Verfaillie provided convincing evidence, contrary to the common view, that mouse AS cells isolated from bone marrow (also known as hematopoietic stem cells) can be stimulated in vitro to differentiate into a wide variety of cell types, representing mesoderm, neuroectoderm, and endoderm, the three fundamental germ layers. The range of cell types representing the ectoderm is more limited than what is possible with ES cells, hence the reference to "neuroectoderm." Consequently, Verfaillie calls her AS cells multipotent adult progenitor cells, or MAPCs. Nevertheless, when MAPCs are injected into a mouse embryo at an early stage of development, they contribute to most, if not all, somatic cell types.

Catherine Verfaillie in her lab at the University of Minnesota in Minneapolis on July 8, 2004. Verfaillie is best known for her work on adult stem cells and is currently at the Catholic University of Leuven in Belgium, where she serves as the director of the Stem Cell Institute.

In recent years, Verfaillie and her group have confirmed and extended their original observations. In 2005, they presented detailed protocols for the growth, stimulation, and maintenance of ES and AS cells. In 2006, they found a way to stimulate the production of smooth muscle cells from MAPCs using just two growth factors. In 2007, the team described a method for reconstituting healthy bone marrow in vivo from isolated MAPCs. And finally, in 2008, Verfaillie's team showed that MAPCs could restore vascular and skeletal muscle growth in mice suffering from limb ischemia due to peripheral vascular disease (i.e., muscle degeneration due to poor blood circulation). Most important, they found that the limbs could also be repaired with human MAPCs.

Blood isolated from human umbilical cords is also known to be an excellent source of adult stem cells. Moreover, these umbilical cord stem (UCS) cells appear to have a developmental plasticity equal to that of ES cells. Dr. Juliet Barker, at the Memorial Sloan-Kettering Cancer Center in New York, has successfully treated leukemia in children using UCS cell therapy. Barker and her associates have also discovered that UCS cells, although allogeneic (i.e., not from the patient being treated), are less inclined to stimulate graft-versus-host disease (GVHD) when compared to allogeneic ES cells. This disease is caused by the immune system, which is programmed to reject foreign cells, tissue, and organs (the grafts). Thus UCS cells offer a great improvement over standard bone marrow transplants, which are eventually rejected even in the presence of immunosuppressants. Many parents are already having umbilical cord blood collected from their newborn infants in case the child should ever need stem cell therapy, and some portion of the blood so collected is being set aside for general use. In time, umbilical cord blood, collected from millions of infants, could become an extremely valuable tissue bank that could be used to treat many diseases without causing immune rejection and without raising the ethical problems associated with ES cells. Currently, there are more than 250,000 units of umbilical cord blood banked for public use, and more than 8,000 UCS cell transplants have been performed.

The existence of AS cells is extremely important, since they resolve the ethical problems that are associated with the use of ES cells. Many people, scientists included, are uncomfortable with the practice of obtaining stem cells from human blastocysts or fetuses and hope that AS cells will eventually resolve the issue. The limited plasticity of AS cells is a major hurdle to overcome before they will be a practical alternative to ES cells, but many researchers believe this can be accomplished.

Even if the social objections to the use of human ES cells did not exist, the study of AS cells would still be essential, since the human ES cells that are now being isolated and cultured are good only for basic research and not for the treatment of human diseases. This is because the ES cells, being allogeneic, will be destroyed by GVHD, the same way the immune system rejects foreign hearts, livers, or kidneys. Organ transplants are common today, but they are not very successful, and the patients must be kept on immunosuppressants for the remainder of their lives. In the case of bone marrow transplants, which are used to fight leukemia, physicians try to find a close genetic match by obtaining the tissue from a parent or sibling. But even in those cases, the match, if one is ever found, is imperfect. Organ and tissue transplants, relying on immunosuppressants, are therapies of desperation. They work well enough to keep the patient alive for a few extra years, but they are not particularly healthy or happy years.

Adult stem cells could solve the problem of GVHD in the following way: AS cells, isolated from a patient's bone marrow, could be grown in culture to increase their numbers before being stimulated to produce the tissue needed to treat that same patient. Stimulation of the AS cells in culture is known as directed differentiation and will be discussed at length in a following section. The use of AS cells removes the ethical problems associated with harvesting human embryos, and it solves the problem of tissue rejection. The cell surface antigens on the tissue so produced would be compatible with the patient's immune system, so there would be no fear of rejection and no need for immunosuppressants. The cost of this procedure is not likely to exceed a standard bone marrow transplant and would take but a few weeks to administer. For the treatment of blood disorders such as leukemia, this is an ideal procedure and is now a routine medical therapy.

Brain disorders can also be treated in this way, since AS cells isolated from bone marrow can be directed to differentiate into neurons and glia cells. However, making this procedure applicable to all forms of disease requires new methods that would enhance the developmental plasticity of the AS cells. Collecting stem cells from bone marrow has many advantages including a reduction in the cost of the procedure. AS cells can be obtained from other organs, such as the pancreas or liver, which would allow production of a wider variety of cell types, but obtaining them would require costly surgeries.

STEM CELL MARKERS

Identifying a stem cell is not always easy. AS cells account for only 1/100,000 of the total cell population, so the odds of finding one, at the best of times, are small indeed. Stem cells have a simple morphology, and one might think this would set them apart from other cells in the body, but there are many differentiated cells that have a similar size and shape. Consequently, it is almost impossible to separate stem cells from differentiated cells by simple visual inspection. The situation with ES cells is much better, since their source and identity are known without question. However, all stem cells change somewhat when placed in culture, making it necessary to monitor their behavior and to track any changes in their state of differentiation.

To this end, a set of markers has been developed that simplify the identification of stem cells and the evaluation of their phenotype. There are many different stem cell markers, but they all fall within one of three groups: glycoprotein receptors that are embedded in the cell membrane, cell-specific gene expression, and cell-specific molecules, such as hormones, enzymes, or structural proteins.

Glycoprotein Receptors

White blood cells carry cell-surface receptors called CD4 and CD8, which are specific for mature T lymphocytes. The protein that binds to these receptors is called a ligand and can be detected with a procedure called immunofluorescence. Briefly, this involves placing the cells on a microscope slide and covering them with a solution containing the ligand. During incubation, the ligand binds to the receptors (if present), after which the sample is covered with a solution containing a fluorescent-labeled antibody that will attach specifically to the ligand. After an appropriate incubation period, the slide is examined under a fluorescent microscope. Cells carrying CD4 or CD8 receptors will be colored blue or green, while all the other cells will be colorless. A negative reaction, where all cells are colorless, indicates the stem cells have not differentiated or they have not differentiated into T lymphocytes.

Fluorescent markers are often used in conjunction with a machine called a fluorescence-activated cell sorter (FACS). A suspension of thousands of cells is treated with the immunofluorescence procedure to tag any stem cells that may be present with a fluorescent dye, after which the sample is injected into the FACS machine. The cell suspension passes through a very thin tube, which forces the cells to move past a laser beam in single file. The laser beam gives each cell an electric charge: Those that carry the fluorescent marker receive a negative charge; the rest receive a positive charge. An electromagnetic field directs the negatively charged cells into one tube, while the positively charged cells are directed to a separate tube. A FACS machine can isolate a single stem cell from a population of more than 100,000 cells in less than an hour.

Cell-Specific Gene Expression

The expression of certain genes in specific cell types is another kind of stem cell marker. Some neurons are known to express a gene called *Noggin*, which is not expressed in nonneural tissue. Using a procedure called fluorescence in-situ hybridization (FISH), it is possible to detect the cells that express the *Noggin* gene. Like the immunofluorescence procedure just described, the final product of the FISH reaction is a field of cells on a microscope slide that are either colored or not. Cells expressing *Noggin* will be blue; those that do not express this gene will be colorless. Stem cells that are differentiating into neural tissue will be colored. Stem cells that are not differentiating, or are differentiating into nonneural tissue, will be colorless.

Cell-Specific Molecules

Some cells produce special hormones, macromolecules, and enzymes that may be used as markers, since they are not found in all cells. The β cells in the pancreas are the only cells in the body that produce insulin, so this hormone makes an excellent marker for the differentiation of a stem cell into a β cell. All cells make a protein called Tubulin, which is an important structural protein, but neurons make their own special neurotubulin, which can serve as a marker of neural differentiation. ES cells produce a protein called Genesis, which plays an important role in gene transcription. ES cells also have special glycoproteins in their membranes, such as stem cell antigen-number 1 (Sca-1) and embryonic antigen-3 (Ea-3). Thus, Genesis, Sca-1, and Ea-3 are markers for the undifferentiated state of a stem cell. The disappearance of these molecules is an early indication the cell is beginning to differentiate, even before an outward change in appearance is evident. Detection of cells carrying these markers is possible using immunofluorescence or FISH. Gene and protein nomenclature is explained in chapter 10. Briefly, gene names are capitalized and italicized, whereas protein names are capitalized and in roman type.

A special DNA sequence called a telomere may also be used in conjunction with the standard markers to help identify a potential stem cell. A telomere is not a gene but a simple repetitive sequence located at the tip of each chromosome that is needed for the proper duplication of each chromosome during the cell cycle. With each round of cell division, the telomere shrinks in length, but it is later restored by an enzyme called telomerase. Actively dividing cells have high levels of telomerase, whereas post-mitotic cells have none. Consequently, the presence of telomerase is an indication that the cells are actively dividing, as stem cells usually are.

Another approach to monitoring a stem cell's differentiation status employs the technique of gene therapy to introduce a reporter gene, coding for a green fluorescent protein (GFP), into the genome of a stem cell. GFP emits a green light and can be engineered so that it is active only when the cell is in an undifferentiated state and is turned off when the cell begins to differentiate. Consequently, differentiation of the cell is associated with a loss of the green color. An alternative use of the *GFP* gene is to use it to give cells a permanent green color. Dr. Fred Gage and his team at the Salk Institute have obtained stunning results on stem cell differentiation by using the *GFP* gene in this way. Their intention was to study the ability of astrocytes (a kind of neuron) to stimulate differentiation of adult stem cells into neurons. The in vitro experiments were conducted using cells isolated from rats. After transfecting stem cells with a *GFP* reporter gene, Gage's team added them to a culture plate containing astrocytes. In this experiment, the reporter gene is always turned on, so *GFP* gave the stem cells a permanent green color. Each time the stem cells divided they produced green daughter cells, so their fate was easily monitored. This procedure gave clear evidence that the astrocytes could induce stem cells to differentiate into neurons, since green neurons appeared that could only have originated from the green stem cells.

Stem cells are generally characterized through the examination of more than one marker and are referred to accordingly. This produces a very specific, though somewhat awkward, naming convention. A neural stem cell (NSC), isolated from the brain and evaluated for the expression of CD8, *Genesis,* and *Noggin* would be referred to as NSC ($CD8^{-/low}$, *Genesis*$^-$, *Noggin*$^+$), meaning the expression of CD8 is absent to low, *Genesis* is absent, and *Noggin* is present. The use of these markers was established in studies involving research animals, such as the mouse, but they are now an indispensable part of the more recent research effort to identify and characterize human stem cells.

DIRECTED DIFFERENTIATION

Collecting human stem cells, whether from an adult or an embryo, is just the first step in a long series of tests and procedures that, it is hoped, will lead to a treatment for a medical disorder. Once the cells are collected, they are grown in culture and stimulated in various ways to determine the types of cells they may produce. This stage of the process is called directed differentiation.

All stem cells have the ability to proliferate in culture and can differentiate into many different kinds of cells. Since the goal is to maintain the stem cells in an embryonic state, initial culturing conditions are chosen to block spontaneous differentiation. This involves growing the cells on a layer of feeder cells that secrete substances into the culture media that help nourish the ES cells. For the most part, the identity of these substances is unknown. The feeder cells are usually mouse embryonic fibroblasts that have been exposed to γ-rays (gamma rays), a form of radioactivity that destroys the cell's ability to replicate.

Feeder cells help maintain the stem cells in an undifferentiated state; they also provide a favorable substrate for the ES cells to grow on. When the stem cells are needed for a directed differentiation experiment, they are transferred to fresh culture plates that lack a feeder cell layer, then given culture media containing one or more growth factors. In some cases, directed differentiation simply involves transferring the cells to fresh plates, lacking a feeder layer, so the cells can form embryoid bodies. Although the embryoid bodies vary with regard to cellular composition, they usually include cells that look like neurons and myocytes.

So far, adult and embryonic stem cells have been stimulated to produce several cell types, either through exposure to various growth factors or by being incubated in mice to produce teratomas. Selecting growth factors for these experiments is an example of educated guesswork. If the intention of the experiment is to produce neurons or epithelial cells, the investigators will select growth factors such as epidermal growth factor (EGF) and nerve growth factor (NGF), both of which are known to influence the proliferation of these cells in vivo. Other growth factors, such as transforming growth factor (TGF) or the hormone insulin, are known to influence tissues derived from the mesoderm, such as muscle and cartilage. In some cases, the growth factors producing a certain kind of cell are unknown. This occurs when stem cells differentiate in vivo or when they are cultured in the presence of fetal bovine serum (blood serum obtained from a fetal cow), which contains many yet-to-be identified growth factors.

Retinoic acid, epidermal growth factor (EGF), bone morpho-genic protein (BMP), and fibroblast growth factor (FGF) are some of the growth factors that have been used on ES cells. All of these factors trigger development of cells that would normally be derived from the ectoderm. Other growth factors, such as activin-A and transforming growth factor (TGF), initiate differentiation of meso-derm-derived cell lines. Hepatocyte growth factor (HGF) and nerve growth factor (NGF) promote differentiation of cells that represent all three germ layers. When these factors are added individually or in combinations to cell cultures derived from embryoid bodies, they give rise to a great variety of cell types.

DIRECTED DIFFERENTIATION OF HUMAN ES CELLS

SOURCE	CONDITIONS	RESULTING CELL TYPES
embryo (H9 cell line)	TGF, EGF, NGF, FGF, BMP, HGF, retinoic acid	neuron, skin, adrenal, blood cell precursors, liver, pancreas, muscle, bone, kidney, heart
embryo (H9 cell line)	teratoma, LIF	bone, cartilage, gut epithelia, neural epithelia, smooth muscle, striated muscle
embryo (H9 cell line)	embryoid body	pancreatic β cells

Note: TGF (transforming growth factor), EGF (epithelial growth factor), NGF (neurotrophic growth factor), FGF (fibroblast growth factor), BMP (bone morpho-genic protein), HGF (hepatocyte growth factor), LIF (leukemia inhibitory factor)

DIRECTED DIFFERENTIATION OF HUMAN ADULT STEM CELLS

SOURCE	CONDITIONS	RESULTING CELL TYPES
bone marrow	teratoma	hepatocyte, myocyte, adipocyte, chondro-cyte, red blood cell, white blood cell, cardiomyocyte
bone marrow	EGF, NGF, FRN	neuron
brain	teratoma, FCS	myocyte, astrocyte, neuron, oligodendrocyte
liver	teratoma	red blood cell, white blood cell

Note: EGF (epithelial growth factor), NGF (neurotrophic growth factor), FRN (fetal rat neuron), FCS (fetal calf serum)

Spontaneous differentiation of ES cells in culture produces several different kinds of cells on a single plate, but stimulating the cultures with any one of the growth factors mentioned above tends to focus the differentiation toward a single cell type. Cultures stimulated with FGF differentiate into epithelial cells that express the marker keratin, a common skin protein. Cultures treated with activin-A produce muscle cells that express a muscle-specific enzyme called enolase. Retinoic acid typically stimulates the production of neurons, but it is also known to initiate development of other cell types.

An important and much-sought-after result of directed differentiation is the production of blood cell precursors for the treatment of leukemia. Here, AS cells are the clear favorites, since they can be stimulated in culture to produce white blood cells and they do not cause GVHD. There is, however, one drawback to using AS cells to treat leukemia: Due to the difficulty of obtaining healthy AS cells from a malignant bone marrow, it is sometimes impossible to treat a patient with these cells. The best alternative is to use the UCS cells discussed above. ES cells could also be used, but producing blood cells from ES cells through directed differentiation has been mostly unsuccessful. Some success has been achieved by growing human ES cells in the presence of γ-irradiated mouse bone marrow cells (γ-irradiation blocks replication of the mouse cells). The mouse cells apparently provide an unknown growth factor that triggers differentiation of the ES cells into blood cells. The differentiated cells express CD34, a marker for blood cell precursors, and under certain conditions these cells will form erythroid cells, macrophages, and other blood cells. The value of this line of research is now in question, however, due to the work of Barker and her associates.

Human ES cells have a greater tendency to differentiate spontaneously when placed in culture than mouse ES cells. Scientists wishing to produce a culture of human myocytes or neurons through directed differentiation must start with a population of undifferentiated cells, otherwise the product of the experiment might be a curious hybrid cell that could yield unpredictable and perhaps fatal results if used in a clinical setting. Markers of the embryonic state, such as stage-specific embryonic antigen (SSEA), are being used, in conjunction with a FACS machine, to isolate and segregate the undifferentiated cells from the rest of the population, so they can be used for directed differentiation. In addition, the isolation of the undifferentiated cells will make it possible to study the differences between those cells that remain embryonic and those that become partially or wholly differentiated.

Directed differentiation of AS cells has confirmed prior in vivo studies that AS cells have less plasticity than ES cells. When stem cells from an adult mouse brain are placed in culture, they differentiate into neural tissue but not the variety of cell types that are produced by ES cells. Likewise, AS cells from the bone marrow differentiate, in vitro, into blood cells, neurons, and fat cells but not glandular tissue. Moreover, scientists have less control over the differentiation process of AS cells than that of ES cells. Adult bone marrow stem cells differentiate spontaneously when placed in culture, and there is no way to stop them. Consequently, it is very difficult to maintain a continuous culture of these cells. So far, only mouse ES cells are able to proliferate in vitro for an indefinite period of time while retaining an embryonic phenotype.

The gradual loss of plasticity in human stem cells may not be a serious problem, since some ES cells have been maintained for more than two years in an undifferentiated state. Moreover, even after months of culturing, stem cells will still respond to directed differentiation. The responsiveness is extremely important if these cells are going to be used to treat a disease. For example, to repair a damaged heart, cultured ES cells, kept in stock for years, would be stimulated to differentiate into myocytes, after which they would be injected into the circulatory system of the patient. The trick is to ensure the cells are only partially differentiated (that is, they are myocyte precursors); otherwise they would lose the power to proliferate and be unable to repair the heart.

Simply injecting freshly isolated ES cells into the patient would lead to the formation of teratomas, as occurs when stem cells are injected into mice. Injected myocyte precursors, on the other hand, will home in to the parent organ, the heart, to initiate repairs. Of course, this scenario is very hypothetical and may contain a large dose of wishful thinking. For the present, scientists simply do not know what a myocyte precursor will do when it reaches the heart and, perhaps more important, no one knows how the heart will respond.

Directed differentiation of cultured human stem cells has led to the production of many cell types, but in most cases scientists do not understand the details of the process. Manipulating the culture conditions, by adding a growth factor or a molecule that influences gene expression, may cause the stem cells to differentiate into nervous tissue or bone, but the intermediate steps that lead to the transformation are still obscure.

Consequently, directed differentiation requires a good deal of guesswork. Various compounds known to influence gene expression can be added to the culture media to see if they will stimulate differentiation and, if they do, which kind of cell is produced. In this way, scientists are in the process of constructing a catalogue of culture additives that will lead to the production of specific cell types from cultured ES and AS cells. The study of directed differentiation may also help scientists understand the molecular nature of plasticity and the factors that distinguish ES and AS cells. This information could pave the way to effective therapies based solely on the use of AS cells.

Growth Factors

The growth factors used for directed differentiation are a fascinating group of molecules. One of the most famous is nerve growth factor (NGF), isolated in 1981 by Dr. Rita Levi-Montalcini after nearly 20 years of trying to track down factors that influence the behavior of neurons during embryonic development. In 1986, Dr. Levi-Montalcini and her collaborator Dr. Stanley Cohen received the Nobel Prize in physiology or medicine for discovering NGF and EGF, an epidermal growth factor used extensively in the treatment of severe burns and in stem cell research. Growth factors are proteins that bind to cell-surface receptors, each of which activates a signaling pathway, leading to the phosphorylation of many cytoplasmic proteins.

Phosphorylating a protein is analogous to flipping a switch to turn on a light. Phosphorylation of a protein turns it on, converting it from an inactive to an active state. When the activated protein is no longer needed, it is turned off by another enzyme, called a phosphatase, which removes the phosphate group. Protein kinase receptors consist of a single polypeptide chain that spans the cell membrane, with the portion inside the cell containing the enzymatic region. The kinase region of the growth factor receptor adds a phosphate group specifically to the amino acid tyrosine. Other protein kinases specialize in adding phosphate groups to serine or threonine amino acid residues.

Protein growth factors bind to their receptor as dimers. That is, two identical growth factor molecules join together to form an activated pair that is capable of binding to the receptor. Binding of the dimerized growth factor then stimulates dimerization of the receptor. Dimerization of the receptor activates its kinase domain, located inside the cell, which then phosphorylates cytoplasmic signaling proteins that stimulate cell growth, differentiation, and proliferation. Each growth factor binds to its own specific receptor, which activates a unique set of signaling molecules. This is the reason that some growth factors function primarily as mitogens (stimulating cell division, but not necessarily differentiation), while others stimulate differentiation as well as proliferation. Identifying the signaling molecules and the pathways they are on is a very active area of research, which is important in many areas of biological research, including cancer and aging research. Stimulating stem cells to grow and differentiate will become a much more precise science when growth factor signaling pathways are understood in greater detail.

Two other growth factors used in directed differentiation are leukemia inhibitory factor (LIF) and retinoic acid. LIF is a cytokine (signaling proteins released by many kinds of cells), originally studied for its ability to force the differentiation of certain kinds of cancer cells. Once they differentiate, the cancer cells lose their ability to grow and divide indefinitely, thus inhibiting the spread of tumors. LIF is also expressed by embryonic trophoblasts and is believed to

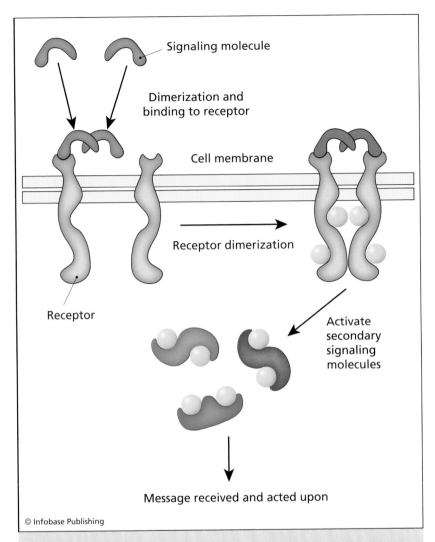

Signaling molecule

Dimerization and
binding to receptor

Cell membrane

Receptor dimerization

Receptor

Activate
secondary
signaling
molecules

Message received and acted upon

© Infobase Publishing

Enzyme-linked receptor. The primary signaling molecule, such as NGF or TGF, forms a dimer (two identical molecules bound together) and binds to the receptor, stimulating dimerization of the receptor. This stimulates phosphokinase activity of the receptor, which phosphorylates itself (orange spheres) and several secondary signaling molecules. The activated secondary molecules deliver the signal to the appropriate part of the cell.

play an important role in the implantation of the blastocyst in the mother's uterus.

Retinoic acid, closely related to vitamin A, is one of the few growth factors that is not a protein. This compound is a popular ingredient in antiwrinkle creams and is taken by some people to improve their night vision. Retinoic acid has been used in many different kinds of experiments to radically alter the normal gene expression profile of a cell. It is a powerful growth factor but somewhat unpredictable in its behavior and effects. Unlike other growth factor receptors, the receptor for retinoic acid is located inside the nucleus. Being fat-soluble, retinoic acid can diffuse directly into a cell and into the nucleus, where it binds to its receptor. The retinoic acid–receptor complex activates gene expression directly without relying on the activation of intermediate signaling molecules.

There are many growth factors yet to be discovered. This is demonstrated by the fact that directed differentiation sometimes fails unless the cells are grown in the presence of other types of cells or in the presence of fetal bovine serum (FBS). FBS is obtained by collecting blood from an aborted cow fetus and spinning it in a centrifuge to remove all of the blood cells. The amber-colored serum, containing many unidentified growth factors and hormones, is used in a wide variety of cell culture experiments. Some cells simply will not grow well unless FBS is added to the culture media. Efforts to determine the identity of the growth factors in FBS are in progress in many labs around the world.

FUTURE PROSPECTS

The speed at which stem cell research is progressing is such that within 10 years AS cells will be used routinely to repair or replace any tissue in the body. ES cell research will decline and eventually disappear entirely. This will come about, in part, through a better understanding of growth factors in order to augment the plasticity of AS cells. Initially, BMS cells or UCS cells will be the starting material for the therapies. But eventually it will be possible to take a simple tissue scrape from inside a patient's mouth to obtain skin cells that will be reprogrammed to become a stem cell and then directed to differentiate into a different kind of cell, depending on what the patient requires. This work has already begun with the production of iPS cells, discussed in chapter 4.

These iPS cells are expected to revolutionize the field of stem cell research and could possibly make ES cells and therapeutic cloning obsolete. Running parallel with these scientific developments is the ongoing ethical debate over the destruction of human embryos in order to harvest their stem cells. This will be discussed at length in chapter 8, but for now it is worth noting that a resolution of this debate is currently nowhere in sight. Part of the problem is due to the fact that human embryos have no legal status and IVF clinics generally produce far more human embryos than are needed for reproductive purposes. The extra embryos are kept frozen, and many scientists believe they should be donated for research. Those opposed to ES cell research insist that extra embryos should be protected and kept in storage until someone adopts them and carries them to term. In any case, this debate is likely to be short-lived if the promise of AS and iPS cells is realized.

Therapeutic Cloning

Isolating stem cells and subjecting them to directed differentiation are only a small part of what must be done to produce an effective medical therapy. There is no doubt that embryonic stem (ES) cells offer a way of curing many intractable diseases, but these cells, deriving from a variety of individuals, cannot be used without the threat of immune rejection and the onset of graft-versus-host disease (GVHD). One solution is to clone the patient, isolate stem cells from the resulting embryos, and then use those cells to treat that same patient. The cloned stem cells, and any cells they give rise to, would carry the patient's unique cell-surface glycocalyx and, consequently, would not be attacked by the immune system. This procedure is called therapeutic cloning.

Needless to say, therapeutic cloning, involving the killing of human embryos, has generated as much controversy as the harvesting of ES cells. Researchers, unwilling or unable to pursue adult stem

(AS) cell research insist that this procedure offers the best chance of successful stem cell therapies. Whether they are right or not remains an open question. This chapter will explore the technical and historical aspects of this form of stem cell therapy, leaving the ethical and legal issues for later chapters.

HISTORICAL BACKGROUND

Cloning a human, as part of therapeutic cloning, involves a technique whereby the nucleus from a skin cell is introduced into an enucleated egg cell (also known as an oocyte). This procedure, known as somatic cell nuclear transfer, depends on the ability of a skin cell nucleus to sustain normal development and the production of a healthy embryo. Scientists know that it can, but only because of experiments that were conducted more than 100 years ago by the great German embryologist Hans Spemann.

In Spemann's time, embryologists were faced with a problem in embryology that seemed irresolvable: Animals begin life as a single cell (the fertilized oocyte) that differentiates into a great variety of cells during the process of embryogenesis. But how can a single cell differentiate into many kinds of cells if all of those cells have the same genes? The underlying mechanism of cellular differentiation was unknown, but there were two possibilities: Genes were either lost or repressed. That is, a differentiated cell either loses the genes that specify other cell types or those genes are turned off or repressed in some way. For example, mature liver cells either lose brain-specific genes or the brain-specific genes are repressed. Similarly, neurons lose or repress liver-specific genes.

Spemann reasoned that if the mechanism for cell differentiation was gene diminution, blastomeres isolated from 2- or 16-cell embryos would be unable to develop into normal adults. On the other hand, if diminution was not the mechanism, a blastomere from a 2-cell embryo should be able to produce a normal embryo and adult. In his first experiment, Spemann used a loop of fine human hair, which he got from his newborn son, to separate a 2-cell salamander embryo into single cells, both of which went on to produce normal salamanders. Spemann concluded from this experiment that blastomeres at the 2-cell stage retain their totipotency.

Next, he decided to extend the experiment by testing the totipotency of older embryos, just in case genetic diminution occurred later in development. To test his hypothesis, Spemann, with infinite patience, tied a noose around a fertilized egg, separating it into a nucleated and enucleated cell. He kept the noose tied until the nucleated cell had divided eight times, at which point he loosened the noose just enough to let one of the nuclei enter the enucleated cell, after which the 16-cell embryo was separated from the 1-cell embryo. Spemann found that development proceeded normally in both embryos and thus concluded that blastomeres from 16-cell embryos were also totipotent. These results represent the first time an animal was cloned experimentally.

Spemann realized that the final test, what he called a "fantastical experiment" would be to transfer a nucleus from a fully differentiated adult cell into an enucleated egg and then see if that nucleus could support normal embryonic development. The technical difficulties involved in carrying out such an experiment were beyond the science of Spemann's day. Consequently, his fantastical experiment was not performed until 1996, when the Scottish embryologist Ian Wilmut cloned a sheep named Dolly. Wilmut used two types of sheep for his cloning experiments: Poll Dorsets, a hornless white sheep, and Scottish Blackface, a horned breed that is black and white. The procedure involved transferring a nucleus from an ovine mammary gland epithelial (OME) cell from a Poll Dorset to an enucleated egg obtained from a Scottish Blackface. In practice, the entire karyoplast (the cell donating the nucleus) is injected into the space between the zona pellucida and egg cell. The karyoplast and the cytoplast (the enucleated egg) are then fused together with an electric current. This is a much gentler procedure, compared to injecting the nucleus into the egg by poking another hole in the egg's

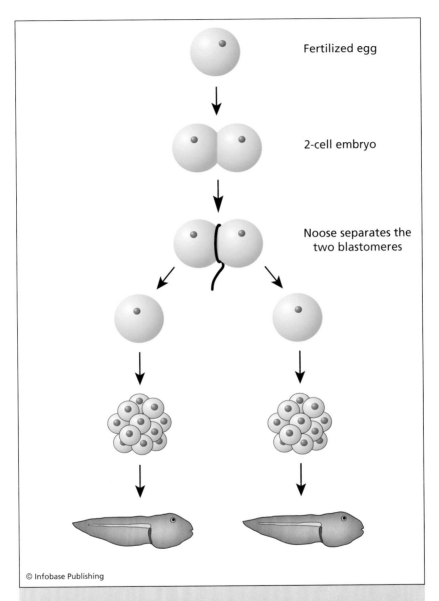

Experiment to test totipotency. A noose was used to separate a two-cell embryo into two blastomeres. Development proceeded normally in both embryos, proving that nuclei at the two-cell stage are totipotent.

Fertilized egg

2-cell embryo

Noose separates the two blastomeres

© Infobase Publishing

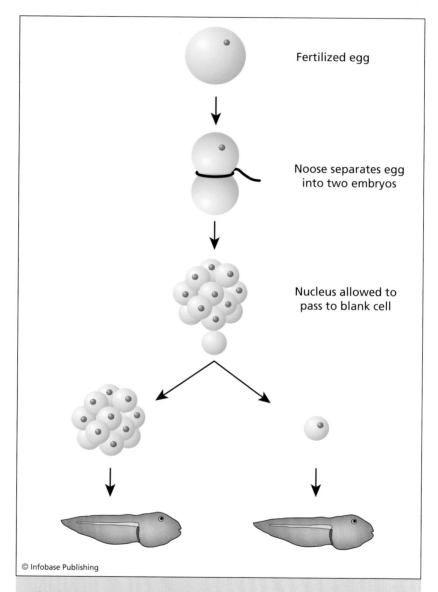

Fertilized egg

Noose separates egg into two embryos

Nucleus allowed to pass to blank cell

© Infobase Publishing

The first cloning experiment. A noose was tied around a fertilized salamander egg to produce one cell with a nucleus and one without (blank cell). At the 16-cell stage, the noose was loosened to allow a nucleus to pass into the blank cell, after which the 16-cell embryo was separated from the one-cell embryo. Development proceeded normally in both embryos.

membrane (the first hole was made when the egg's chromosomes were removed). Once development begins, the embryo is encased in agar and incubated temporarily in the oviduct of a Blackface, after which it is transferred to the oviduct of the final surrogate mother, also a Blackface, which carries it to term.

Wilmut and his team constructed a total of 277 embryos, of which 29 developed to the blastocyst stage in the temporary surrogate. These blastocysts were removed from the agar and transferred to 13 Blackface ewes, one of which became pregnant and gave birth to Dolly. With only one out of 277 embryos going full term, it is a wonder the experiment worked at all. Wilmut could tell at a glance that the experiment had worked, since Dolly was clearly a white-faced Poll Dorset and not a Scottish Blackface. Despite the obvious differences between the surrogate mother and Dolly, extensive DNA tests were conducted to prove that Dolly's genome came from the culture of OME cells and that she was indeed a clone of one of those cells.

CLONING HUMANS

The ability to clone a human has so far evaded all attempts, even though the basic procedure was described more than 10 years ago when Dolly the sheep was cloned. Part of the reason for this is due to the extremely low efficiency of somatic cell nuclear transfer, with only one in 277 sheep embryos going to term. Cloning efficiency of farm animals has improved over the years but is still less than 1 percent. There are bound to be some slight differences between farm animals and humans with regard to this procedure, which could easily mark the difference between success and failure.

Another reason for the difficulty in obtaining human clones is almost certainly due to the ethical controversy that has limited funding for this kind of research and the availability of human oocytes. In the United States, researchers cannot by law offer to pay women for their eggs although they can accept donations. Never-

The Scottish scientist Ian Wilmut. Wilmut is seen in St. Paul's Church, Frankfurt, Germany, in 2005, prior to being awarded the Paul Ehrlich and Ludwig Darmstaedter Prize. (*AP Photo/Michael Probst*)

theless, for nearly eight years researchers in Europe and the United States tried but failed to obtain cloned human embryos. It was thus a shock to the research community when in 2004 a research team in South Korea, under the leadership of Dr. Hwang Woo Suk, announced that they had succeeded in cloning human embryos and had produced patient-specific embryonic stem cells.

STEM CELL FRAUD: THE CASE OF HWANG WOO SUK

Dr. Hwang Woo Suk was trained as a veterinarian and worked in that field for a brief period before joining the faculty at Seoul National University (SNU), where, it is said, he worked diligently from 6:00 A.M. to midnight, seven days a week. Between 1999 and 2003, Hwang announced the cloning of several dairy cows, one of which was supposed to be resistant to bovine spongiform encephalopathy (BSE, also known as mad-cow disease). These accomplishments were accepted by SNU and the local science community even

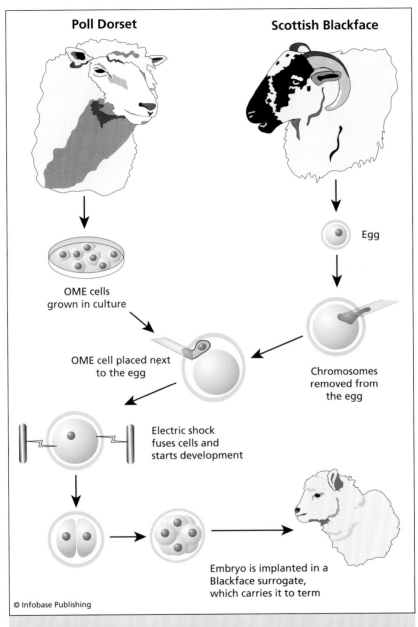

Poll Dorset

Scottish Blackface

Egg

OME cells
grown in culture

OME cell placed next
to the egg

Chromosomes
removed from
the egg

Electric shock
fuses cells and
starts development

Embryo is implanted in a
Blackface surrogate,
which carries it to term

© Infobase Publishing

Cloning sheep. The Poll Dorset provides the nucleus, which is obtained from cultured ovine mammary gland epithelial (OME) cells. The Blackface provides the egg, which is subsequently enucleated. If the cloning process is successful, the clone will look like a Poll Dorset.

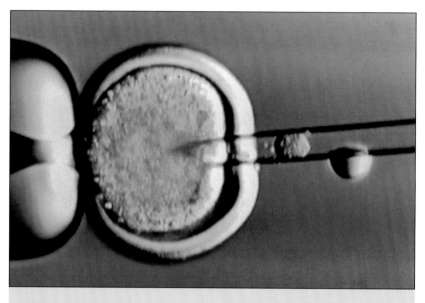

Light micrograph of a sheep egg being injected with an embryonic cell during sheep cloning. The egg (at center) has had its DNA genetic material removed. At left a pipette holds the egg; at right a microneedle injects an embryonic sheep cell into it. (*James King-Holmes/Photo Researchers, Inc.*)

though Hwang made no attempt to prove his claim, nor was his work ever published in peer-reviewed journals.

In February 2004, Hwang announced the successful cloning of human embryos and in 2005 reported the establishment of 11 patient-specific human ES cell lines. Both accomplishments were published in the journal *Science.* Hwang became an overnight sensation. The Korean government made him the head of the World Stem Cell Hub, his work was profiled in *Time* magazine as one of the most influential people of 2004, and famous scientists from the West, including Ian Wilmut, visited his lab to bask in the reflected glory. But in November 2005, Gerald Schatten, a stem cell researcher at the University of Pittsburgh and coauthor on Hwang's papers,

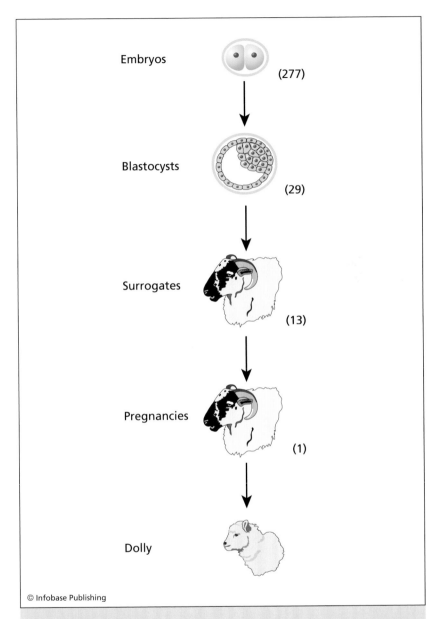

Embryos (277)

Blastocysts (29)

Surrogates (13)

Pregnancies (1)

Dolly

© Infobase Publishing

The Dolly experiment. The difficulty of reprogramming adult nuclei is apparent from the very large number of embryos produced, from which only a single lamb, Dolly, was born.

Dolly the sheep at seven months. Dolly was euthanized at six years of age after developing a progressive lung disease. Her premature death has fueled the debate over cloned animals. (*AP Photo/PA/Files*)

announced that he was no longer collaborating with Hwang and his team. Young researchers in Hwang's lab had contacted Schatten to warn him that the research was strewn with irregularities. Initially, the focus of attention was on the human eggs that were used to generate the clones. Hwang had claimed that they were donated, but in fact women in Hwang's lab were coerced into donating their eggs, while many other women were paid as much $1,400 for their eggs.

The situation quickly became worse when the validity of the research itself was questioned. SNU launched an investigation, which concluded in January 2006 that all of the results published in *Science* were fraudulent. Before the end of that month, Hwang had accepted the blame for the fraud, was stripped of his honors at the university, and both papers were retracted. In February 2006, Hwang's home was raided, and he was fired from SNU. In May of that year, he was indicted on criminal charges, which included fraud, embezzlement, and violation of ethics laws. The trial is primarily concerned with the millions of dollars that Hwang embezzled, most of which came from wealthy donors and the South Korean government. Hwang used some of the money to buy expensive gifts for his family, for some of his sponsors, and for powerful politicians. It was during this trial that Hwang admitted forging data and paying 25 women more than $34,000 for their eggs. It was also established that Hwang and some of his associates had embezzled millions of dollars by inflating research-related expenses. The trial was concluded in October 2009: Hwang was convicted of embezzlement, research fraud, and illegally buying human embryos. He was given a two-year suspended sentence.

FUTURE PROSPECTS

The future of therapeutic cloning is bleak. Aside from the ethical problems associated with human ES cell research and the acquisition of human eggs, no one to date has been able to produce patient-specific ES cells with this procedure. The final nail in the coffin came in 2007 with the production of induced pluripotent stem (iPS) cells (discussed in the next chapter), a type of stem cell that is patient-specific, relatively easy to produce, and does not require human eggs or the killing of human embryos.

Many scientists still believe therapeutic cloning has merit, but Ian Wilmut, inventor of the core technology, is not among them. In November 2007, Dr. Wilmut announced that he was giving up on this procedure in favor of iPS cells and is currently in charge of a cellular reprogramming group at the University of Edinburgh.

Stem Cell Induction

In November 2007, a team of Japanese scientists led by Dr. Shinya Yamanaka announced to the world that they had succeeded in converting a humble skin cell into something that is equivalent to an embryonic stem (ES) cell. Scientists call this cell an induced pluripotent stem (iPS) cell.

Although this accomplishment may seem as though it came out of the blue, the theoretical aspects of the procedure had been on scientists' minds since the 1800s. Even with a rudimentary knowledge of cellular genes, it is possible in theory to convert a muscle cell into a neuron or a liver cell into a skin cell. It is just a matter of tinkering with the cell's genome. Knowing which genes to turn on or off and gaining the ability to do so unleash the full power of biotechnology while opening the door to unlimited possibilities. Of course, nature has been playing this game for a very long time—its experiments with altering the identity of a differentiated cell are formally known

Shinya Yamanaka. Dr. Yamanaka is the Kyoto University scientist who invented the technology to create iPS cells from ordinary tissue like skin. (AP Photo/Shizuo Kambayashi)

as cellular transformations. In plain language, it is known as cancer: the natural conversion of one type of cell into another, only in this case the product is something that nobody wants and everyone dreads. Fortunately, the creation of a stem cell by manipulating the genes of a skin cell is much more benign and holds tremendous potential.

HISTORICAL BACKGROUND

The technology that was used to produce iPS cells began to form in the 1800s when Hans Spemann showed that embryonic development is not associated with a loss of genes. Every cell has a full genome, but some genes are repressed. That is, liver and skin genes are repressed in neurons, and likewise neuron genes are repressed in liver and skin. Spemann's experiment not only resolved an important puzzle, it also revealed something very interesting about cells: They have the ability to turn genes on and off. An oocyte has a simple spherical morphology and does not look at all like a neuron, a skin cell, or a liver cell. Thus, the genes that specify those cells are turned off in an oocyte and are turned on at the appropriate time during development. The mechanism underlying this remarkable process was completely unknown in Spemann's time. Indeed, scientists of that era knew nothing about the structure of a gene and could not even say which type of molecule a gene was made from.

This began to change in 1952, when Martha Chase and Alfred Hershey proved that deoxyribonucleic acid (DNA) is the genetic material of a cell. In 1953, Francis Crick and James Watson published their now classic paper entitled "A Structure for Deoxyribose Nucleic Acid," in which they described the molecular structure of DNA. Their paper not only proposed a structural model for the DNA molecule but also showed how DNA could store a genetic code, specifying a unique protein, and how that code could be duplicated, in a process now known as DNA replication. For the first time, a gene became a concrete reality: a string of four different nucleotides, linked together like beads on a string, that formed a code specifying a single protein. (For more information, see chapter 10.)

All cells are biochemical entities that rely primarily on a large army of proteins to regulate cellular activities. Proteins also provide the structural materials to build and repair the cell's outer protective membrane and its internal compartments. Scientists know that DNA encodes the information needed to make the proteins, but the protein-synthesizing machinery is located in the cytoplasm, while the DNA is located in the nucleus. How is the information relayed to the cytoplasm? Watson and Crick were the first to answer this question by proposing the existence of a molecular intermediary, called messenger ribonucleic acid (mRNA), between DNA and protein synthesis. Ribonucleic acid (RNA) is a close relative of DNA. In a real-world office analogy, mRNA would be a photocopy of a blueprint. The photocopy can then be given to carpenters or engineers to make whatever is required: a chair, a boat, or a house. Thus, the mRNA, like any photocopy, is expendable while the original code or blueprint is kept safe and out of harm's way. The basic flow of command that occurs in all cells was now clear: DNA, containing all the blueprints, is copied into mRNA, which is then used to guide the synthesis of a specific protein. Copying DNA into mRNA is known as transcription, and the second step, the synthesis of the protein, is called translation. Transcription is carried out by a protein called RNA polymerase. Translation is performed by a multienzyme complex known as a ribosome. These steps were first identified in bacteria, an ancient form of cellular life belonging to a taxonomic group called the prokaryotes.

Having arrived at this eloquent flow of command, scientists realized there had to be some way for the cell to know which part of the DNA was to be transcribed. In addition, unless the gene was always active (i.e., always turned on), there had to be some way to turn it on and, when it was not needed anymore, to turn it off. When a gene is transcribed, it is said to have been expressed, and thus the process is known as gene expression. The study of this topic began in earnest in the 1960s when scientists began

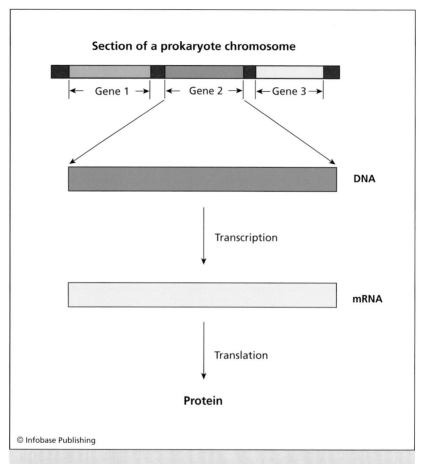

Prokaryote genes. The genes are arranged in tandem along the chromosome, with little if any DNA separating one gene from the other. The genes may code for protein, as shown above for gene 2, or ribosomal RNA (rRNA).

examining the control of gene expression in bacteria. This work continued throughout the 1970s and 1980s, coinciding with the invention of biotechnology.

It turns out that all genes are equipped with a controlling region called the promoter that serves the same function as a light switch and provides the binding site for the RNA polymerase. The switching function of the promoter is regulated by other proteins, known as transcription factors. In some cases, two or more genes are controlled by the same promoter, so when the promoter switches on, all the genes are activated simultaneously (i.e., transcribed into RNA). This arrangement, in which a single promoter is linked to two or more genes, is known as an operon, and it is very common among prokaryotes. Indeed, it was the study of the bacterial *Lac* operon that gave scientists their first insights into the control of gene expression. The *Lac* operon has three structural genes that code for proteins needed for the import and processing of the sugar lactose, from which this operon gets its name. By convention, these genes are known as *Z, Y,* and *A.* The *Z* gene codes for β-galactosidase, an enzyme that hydrolyzes lactose to galactose and glucose. The *Y* gene codes for a permease that facilitates the entry of lactose into the bacteria, and the *A* gene codes for a transacetylase that metabolizes lactose-like compounds.

A repressor protein binds to the promoter in a region called the operator, thus preventing the binding of the polymerase. This happens whenever glucose, the preferred sugar, is available. As long as the repressor is bound to the operator, the operon is turned off. Depletion of glucose stimulates the binding of a protein called catabolite activator protein (CAP) to another area of the promoter called the regulator, or CAP site. If lactose happens to be present when the glucose is depleted, the repressor is forced off the promoter, allowing the binding of the polymerase. Once bound, CAP activates the polymerase and the *Lac* operon is transcribed.

The control of gene expression in eukaryotes (i.e., plants, fungi, and animals) is more complex, but the basic logic remains the same. The polymerase, along with a core team of transcription factors, assembles at a promoter. Initiation of transcription occurs when a gene-specific (GS) transcription factor binds to a DNA sequence, known as an enhancer (or enhancer element), located some distance from the promoter. The enhancer is roughly

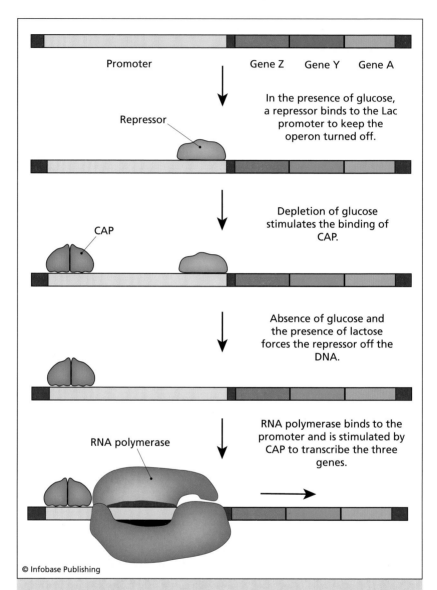

The *Lac* operon. This operon consists of three structural genes (*Z, Y,* and *A*) under the control of a single promoter. The repressor binds to a portion of the promoter called the operator. The catabolite activator protein (CAP) binds at the opposite end of the promoter in a region known as the regulator, or CAP site. Binding of CAP displaces the repressor, thus allowing the binding of the handlike RNA polymerase. The *Lac* operon is designed to stay off as long as glucose is present.

equivalent to the CAP site, or regulator, in prokaryotes; only in eukaryotes the enhancer is located farther away from the promoter. GS transcription factors belong to several families of gene regulatory proteins, the most important of which are the POU, helix-loop-helix (HLH), and zinc finger proteins. POU and HLH factors are similar in the sense that both rely on alpha protein helices, oriented in a precise fashion, to interact with the DNA. Zinc finger proteins have four fingerlike projections, held together by zinc ions, that interact directly with the DNA. Gene regulatory proteins often work by forming homodimers (two identical protein strands linked together) or heterodimers (two nonidentical strands). GS transcription factors play a crucial role in the production of iPS cells.

CREATING PLURIPOTENT STEM CELLS

Yamanaka and his team reprogrammed skin cells by infecting them with viruses carrying just four GS transcription factors. They did not simply guess at which transcription factors would produce stem cells. Rather, they painstakingly tested 25 known factors in various combinations before they found the winning team, consisting of Oct-4, Sox2, Klf4, and c-Myc. (Gene and protein nomenclature is discussed in chapter 10.) By convention, gene names are italicized, but the protein product is not. For example, the Oct-4 protein is encoded by the *Oct-4* gene.

Oct-4

This gene regulatory protein, also known as Oct-3 or Octamer-3/4, was named for the fact that the enhancer element it binds to is eight nucleotides long (ATGCAAAT). This GS factor, belonging to the POU family, has been studied extensively in developing embryos where its expression is confined to undifferentiated, toti-, or pluripotent cells. The Oct-4 protein is known to be present in the nuclei of 8-cell mouse embryos and all cells in the morula stage. At the blastocyst stage, *Oct-4* expression is turned off in the trophoblast but is maintained in the inner cell mass (ICM), which consists entirely of embryonic cells.

In vitro, *Oct-4* is expressed in undifferentiated ES cells and is turned off when they are induced to differentiate by retinoic acid. In studies such as these, confirmation of a cell's status (i.e., whether it is embryonic or differentiating) is established by testing it for the presence of stem cell markers, such as SSEA-1, by its ability to form teratomas, by the types of mRNAs expressed, and by the ability to form healthy mice when introduced into an embryo at the blastocyst stage.

Oct-4 is thought to regulate many genes, but only a few of these have been identified, none of which seem to have a direct role in maintaining pluripotency. Thus, it is unclear how this transcription activator exerts its effect. One suggestion is that it interferes with genes that are known to induce differentiation. Although it is clear that *Oct-4* expression is necessary to maintain the pluripotency of cells, the results of several studies have shown that it does not act alone, but requires the assistance of the other three transcription factors.

Sox2

This transcription activator was originally studied in the context of sexual determination during the development of *Drosophila* embryos and thus its name is an acronym for "sex determination Y-box 2." Scientists now know that its main job is to help embryonic cells maintain their pluripotency. In later stages of development, it appears to have an important role in the differentiation of the nervous system. The expression of *Sox2* is invariably associated with cells that have retained the ability to divide and is turned off in those cells, such as neurons, that have become post-mitotic (i.e., can no longer divide).

It is not clear how Sox2 augments the behavior of Oct-4. It is possible that it forms a heterodimer with Oct-4 to activate essential, though currently unidentified, stem cell genes. Others believe that

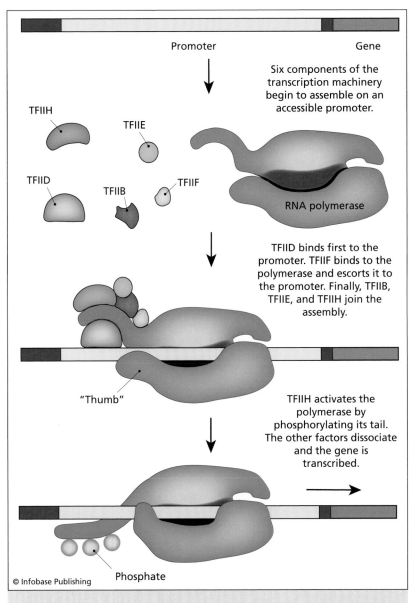

Promoter   Gene

Six components of the transcription machinery begin to assemble on an accessible promoter.

TFIIH

TFIIE

TFIID

TFIIB   TFIIF

RNA polymerase

TFIID binds first to the promoter. TFIIF binds to the polymerase and escorts it to the promoter. Finally, TFIIB, TFIIE, and TFIIH join the assembly.

"Thumb"

TFIIH activates the polymerase by phosphorylating its tail. The other factors dissociate and the gene is transcribed.

© Infobase Publishing   Phosphate

The control of eukaryote gene expression. RNA polymerase and five core transcription factors assemble on an exposed promoter. Once these components are in place, transcription factor IIH (TFIIH) is activated by a gene-specific transcription factor, analogous to the prokaryote CAP. Notice that the "thumb" gets a better grip on the DNA before the gene is transcribed (bottom).

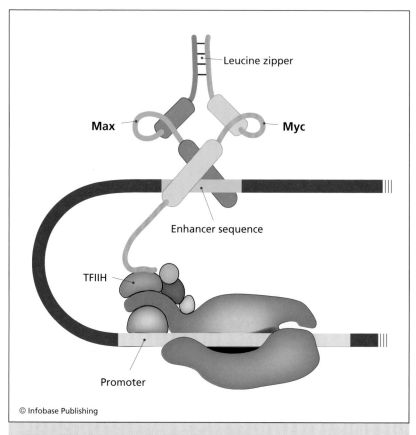

© Infobase Publishing

The activation of a eukaryote gene. TFIIH is activated by a gene-specific transcription factor that binds to an enhancer sequence. The example shown is Myc, which forms a heterodimer with a protein called Max. Myc and Max are held together by four chemical bonds known as the leucine zipper. Myc is a member of the helix-loop-helix family of transcription factors (the helices are shown as barrel structures).

some stem cells genes may require activation through two enhancer elements, one occupied by Oct-4 and the other by Sox2.

Klf4

Klf4, also known as Kruppel-like factor 4, is a zinc finger regulatory protein that was originally identified by *Drosophila* geneticists as having a role in the segmentation of the larvae during embryonic development. In 2006, researchers discovered that Klf4 cooperates with Oct-4 and Sox2 to maintain cells in an embryonic state. The exact mechanism by which this occurs is unknown, but it appears that Klf4 has a physiological role in the activation of *Oct-4* and *Sox2* target genes in ES cells.

C-Myc

This regulatory protein (pronounced c-mick) was isolated in the 1980s from patients suffering from a cancer called Burkitt's lymphoma, a form of leukemia that is technically known as a myelocytomatosis, from which the protein got its name. The *c* refers to the fact that it was isolated from cells to distinguish it from a closely related form that is isolated from viruses (v-Myc).

C-Myc normally stimulates cells to grow and divide. But if the *c-Myc* gene is mutated, the cell loses its ability to regulate the production and activity of c-Myc, and when that happens, the mutant protein forces the cell to proliferate uncontrollably, leading to the formation of cancer. In the mutated state, c-Myc is called an oncoprotein (or cancer-causing protein), and its gene is called an oncogene. This factor's role in maintaining cellular pluripotency is almost certainly due to its ability to stimulate proliferation, since growth and division are the hallmarks of embryonic cells.

DIRECTED DIFFERENTIATION

Initially, there was some concern that iPS cells might not respond to directed differentiation. This seemed like a reasonable concern since the added transcription factors might have been too powerful to overcome, thus forcing the cells to remain undifferentiated. Fortunately, this has not occurred, and indeed iPS cells have responded extremely well to directed differentiation. Within weeks of Yamanaka's announcement, researchers around the world began making iPS cells and testing their potential. The following is a summary of the most important studies.

In 2007, Rudolf Jaenisch at the Whitehead Institute for Bio-medical Research in Cambridge, Massachusetts, used iPS cells to treat and cure sickle-cell anemia in mice. In April 2008, Jaenisch's team produced iPS cells from rats suffering from Parkinson's disease. The cells were differentiated into neural precursor cells that went on to form neurons and glia cells after being injected into the brains of fetal mice. This step validated the embryo-like character of the iPS cells. Finally, the neural precursor cells were differentiated into dopamine neurons and then injected into the brains of adult rats suffering from Parkinson's disease. This treatment was not a cure, but the researchers noticed a marked improvement in the rats' motor skills.

In July 2008, Kevin Eggan and his associates at the Harvard Stem Cell Institute in Boston produced iPS cells from two elderly patients suffering from amyotrophic lateral sclerosis (ALS). This is a neurological disease, also known as Lou Gehrig's disease, that is characterized by a gradual degeneration of motor neurons. Eggan's team induced the iPS cells to differentiate into motor neurons, and the team is confident that with some refinements of the procedure the cells could be used to treat ALS.

In 2009, Kristin Baldwin's team at the Scripps Research Institute in La Jolla, California, pushed the procedure to its logical conclusion by using mouse iPS cells to produce viable embryos that developed into healthy adults. Thus, iPS cells have passed the most stringent test of pluripotency and are thus functionally equivalent to ES cells.

FUTURE PROSPECTS

Ever since the day Yamanaka announced the production of iPS cells, scientists have been asking the big questions: Is this the end of therapeutic cloning? Should everyone stop working on human ES cells? Was then President Bush right all along?

Needless to say, the answers that one can get to these questions may depend a lot on one's political affiliation or religious persuasion. Politicians, members of the clergy, and the general public all seem relieved that iPS cells have come along, that it might mean the end of human ES cell research and the end of controversy. Scientists, on the other hand, are almost unanimous in their desire to continue working on human ES cells. They insist that there is still much that needs to be learned about totipotency and that iPS cells, as good as they seem to be, cannot teach it. They also point out that iPS cells are created by infecting the cells with a virus, which could lead to cancer induction.

But these objections have a hollow ring. Basic properties of ES cells can be learned from mouse, rat, or primate stem cells. After all, the bulk of human genetics was learned by studying bacteria and the humble fruit fly. The issue of viral cancer induction is real, but resolving this problem is insignificant compared to the problem of overcoming graft-versus-host disease (GVHD), a disease that would occur every time a physician tried to treat a patient with human ES cells. Indeed, Yamanaka's team has already produced iPS cells without using *C-Myc,* and other researchers have produced iPS cells without using viral vectors.

The reluctance of the science community to let go of human ES cell research sounds like an empire builder's lament: So much time and energy has been put into it, so much money has been spent on it, so many careers have depended on it—we cannot let it go. But whether scientists want to admit it or not, iPS cells, with their infinite potential, have already made therapeutic cloning and human ES cell research obsolete.

Medical Applications

Scientists were originally interested in stem cells because they provided a way to study totipotency and the plasticity of cells during embryonic development. But it is clear now that these cells can be used to treat, and possibly cure, a wide variety of diseases. Many of the diseases described in this chapter are also being treated with gene therapy, which attempts to correct a genetic abnormality by introducing a normal copy of the affected gene into appropriate cells in the body. This an extremely powerful therapy, but the use of viral vectors to deliver the good gene can sometimes lead to deadly consequences. Stem cell therapy, on the other hand, does not depend on viral vectors but attempts to treat a disease by introducing whole human cells into the body, which, it is hoped, will restore the patient's health. The exception is iPS cells, stem cells produced using gene therapy.

Both therapies are in their early stages of development, so it is not possible to say which is the better therapy. For now, it appears that some diseases may best be treated with gene therapy, while others may respond better to stem cell therapy. Diseases affecting blood cells, such as leukemia or adenosine deaminase (ADA) deficiency, or damage to the spinal cord are best treated with stem cell transplants, whereas cancers of solid organs, such as the brain or lungs, are best treated with gene therapy. Of the two kinds of therapies, treatment with stem cells is more labor intensive. The cells must be isolated, stimulated to differentiate into the desired cell, grown in culture, and then injected into the patient. A major problem associated with stem cell therapy is immune rejection. If the therapy is allogeneic (meaning that stem cells are isolated from an individual other than the patient), the stem cells will be attacked and destroyed by the patient's immune system long before they have a chance to cure the patient's disease. Immune rejection is a major problem to overcome and is discussed at length in chapter 7. Immune rejection is also a serious problem for gene therapy.

The diseases described here represent those cases where stem cell therapy has a chance of being effective or where clinical trials have already demonstrated the effectiveness of the treatment. Much of the work with stem cells is preclinical, relying on results obtained from mice or rats. In these cases, mainly neurological disorders, Phase I clinical trials are still several years away. (The organization of clinical trials is described in chapter 10.)

CARDIOVASCULAR DISEASE

Cardiovascular disease (CVD) often leads to a cardiac infarction (heart attack), which is caused by a chronic reduction in the blood supply to the heart. CVD is the primary cause of death throughout the world. Serious obstruction of the coronary arteries leads to the death of cardiac muscle, the cells of which are known as cardiomyocytes. Depending on the extent of the damage, the heart could be weakened, but still functional, or it could fail completely. At present, the only treatment for cardiac failure is an organ transplant, but over the past seven years scientists have been trying to develop an alternative cell-based therapy.

One of the earliest studies that attempted to treat cardiac failure with stem cells was conducted in 2001 by Piero Anversa's team at the New York Medical College and researchers at the National Institutes of Health (NIH). Adult stem (AS) cells were isolated from the bone marrow of mice and injected into the heart of animals that had suffered an experimentally induced heart attack. The researchers found that newly formed myocardium occupied 68 percent of the damaged portion of the ventricle nine days after transplanting the bone marrow cells. The developing tissue appeared to consist of proliferating cardiomyocytes and vascular structures.

These results led quickly to the initiation of more than 30 CVD Phase I clinical trials with a total enrollment of more than 1,700 patients. Most of these trials were concluded by 2008, but the results were not encouraging. Patients who received injections of their own bone marrow–derived AS cells showed very modest, if any, improvement in cardiac function. This was determined by measuring the fraction of blood within the left ventricle ejected during one contraction, a measurement known as the ejection fraction. Usually, the ejection fraction did not exceed 5 percent, the minimum change required for long-term improvement of symptoms and survival. There was, however, some improvement in exercise capacity, which was an encouraging sign for the treated group.

In an attempt to understand these disappointing results, scientists have gone back to reexamine all of these studies, including the preclinical study of 2001. To begin with, there is now a greater appreciation for the type of environment that the transplanted cells have to deal with. After a heart attack, macrophages remove damaged cardiomyocytes and replace them with scar tissue, which is surrounded by weakened and poorly contractile survivors. In addition, the heart attack itself puts the immune system on alert, so that most of the transplanted cells, even though they are autogeneic (i.e., come from the patient), are likely killed by the macrophages as they try to clean up the damaged area. Those that remain are left with the seemingly insurmountable job of displacing the scar tissue so they can integrate with the healthy muscle tissue. Final integration with the heart tissue requires that the transplanted cells begin to beat at the same frequency as the native cardiomyocytes (60 to 100 beats per minute). This latter requirement is likely the reason a number of cross-species preclinical trials have failed (not discussed here). Such trials involve the transplantation of human embryonic stem (ES) cells into a damaged mouse heart. Cardiomyocytes derived from human ES cells beat at the rate stated above, but the rate of the host mouse heart is nearly 300 beats per minute.

Although the preclinical trial of 2001 concluded that the transplanted AS cells had differentiated into cardiomyocytes, there is now some doubt regarding this conclusion. The researchers tried to confirm the identity of the transplanted cells after they had time to differentiate by looking for the expression of the cardiac transcription factors Nkx-2 and Islet-1. The transplants did begin expressing these factors, but it was wrong to conclude that the cells had differentiated into fully functional cardiomyocytes. Scientists now believe that most, if not all, of the AS cells failed to differentiate into cardiomyocytes. The observed improvement in the function of the mouse hearts was due to the secretion of growth factors and other substances from the AS cells. This is also thought to explain the poor results of the clinical trials described above. The injected AS cells failed to differentiate into functional cardiomyocytes, and the small improvements that were observed are likely due to factors released by the transplants.

Some scientists have argued that the results of these trials would have been much improved if the researchers had used ES cells instead of AS cells. But aside from the ethical problems, ES cells have

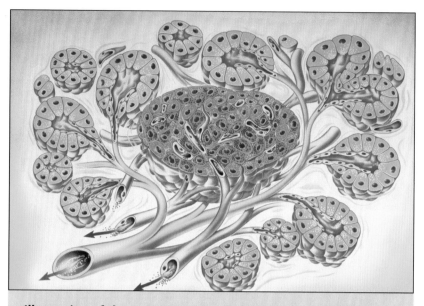

**Illustration of the secretory cells of the human pancreas, show-
ing an islet of Langerhans (center) surrounded by pancreatic acini.** The pancreatic acini produce digestive enzymes (white spots) that
pass through ducts (yellow) into the small intestine. The cells of the
islet are involved in the endocrine function of hormone production.
The hormones (blue and white spots) are secreted directly into the
bloodstream. These hormones include insulin (blue spots, produced
by beta cells), which lowers blood sugar levels, and glucagon (white
spots, produced by alpha cells), which raises blood sugar.
(*John Bavosi/Photo Researchers, Inc.*)

their own serious disadvantages. Two of the greatest problems are the
production of teratomas and the induction of graft-versus-host disease
(GVHD), since the ES cells would all be allogeneic (i.e., not from the
patient). ES cells, once injected into the heart, might differentiate into
cardiomyocytes and integrate fully with the resident cells, but some of
them would travel elsewhere to form teratomas, and all of them would
illicit an attack from the immune system. These problems were taken
into account when the FDA approved the 30 trials discussed above,
and they are the reason ES cells were not allowed.

The lessons learned from the clinical trials for heart disease are so important that all stem cells researchers are studying these trials in the hope that the information will improve cell-based therapies for any disease, whether it be a heart attack, diabetes, or a neurological disorder. Analysis of the CVD clinical trials has identified several problems that need to be addressed: 1) Scientists need better criteria for determining the fate of the transplanted cells; 2) formation of tera-tomas must be avoided at all costs; if induced pluripotent stem (iPS) or ES cells are used, they must be induced to a partially differentiated state before being injected into a patient; 3) more precise methods are needed for directed differentiation in order to produce very homogeneous cell populations; and 4) the stem cells prepared for the therapy must be carefully screened to remove all undifferentiated cells.

DIABETES

Diabetes is a chronic metabolic disorder that destroys the body's ability to utilize glucose, a molecule that is critically important to all cells, particularly neurons, as an energy source. The uptake of glucose is regulated by a hormone, called insulin, that is produced by the pancreas, a large gland located just below the liver. The pancreas has two types of cells, called α (alpha) and β (beta). The α cells produce digestive enzymes that are secreted directly into the large intestine. The β cells produce insulin, which stimulates the uptake of glucose by all the cells in the body. When diabetes strikes, β cells lose the ability to manufacture and release insulin, leading to a buildup of glucose in the blood. A chronic elevation of blood glucose levels results in the inappropriate glycosylation (addition of sugar to proteins) of many proteins in the blood, including hemoglobin, as well as many other proteins associated with the cells and tissues. Systemwide protein glycosylation can lead to blindness, heart disease, kidney failure, and neurological disease. Diabetes is a major health problem in North America, where it causes approximately 500,000 deaths every year. Treatment is very expensive, amounting to about $98 billion annually.

There are two forms of diabetes—type 1 and type 2. Type 1 diabetes is an autoimmune disease in which the white blood cells attack and destroy the β cells of the pancreas. This form of the disease is sometimes called juvenile diabetes because it occurs predominately in teenagers, although it can strike at any age. Type 2 diabetes affects older people, usually when they are 50 to 60 years of age. In this case, the disease may be due to a genetic predisposition to short-lived β cells or it may be due to β cell burnout, brought on by a lifelong preference for a diet that is heavy on sweets. This may account for the fact that nearly 80 percent of those suffering from type 2 diabetes are overweight. At last count, 10 genetic loci were known to be associated with the onset of both types of diabetes.

Diabetes is currently treated with daily injections of purified insulin. While this treatment controls the immediate danger of high blood glucose levels, it does not cure diabetes or remove the long-term threat of kidney failure or the other complications that are associated with this disease. For this reason, scientists have studied the possibility of curing this disease with stem cells. Research teams in Israel and at the NIH in the United States have found a way to direct the differentiation of cultured ES cells into β pancreatic cells that secrete insulin. Other investigators are trying to repeat this accomplishment using AS and induced pluripotent stem (iPS) cells. If this can be done, stems cells could be harvested from each diabetic patient, stimulated to differentiate into β cells, and then returned to the patient in the hope that they will colonize the pancreas, thus curing the disease.

Scientists at the biotechnology company Novocell recently tested this procedure in mice. They began by turning human ES cells into β cell precursors and then injected them into mice. The cells did not colonize the pancreas, but they did produce insulin and appeared to respond to normal physiological cues. An important lesson to be learned from this study is that the transplanted cells did not have to integrate within the pancreas to be effective, thus simplifying the complexity of the procedure. Unfortunately, about 7 percent of the mice developed teratomas, and it is for this reason that this research has not yet progressed to a clinical trial. It simply makes no sense to risk giving a patient cancer, which may not be treatable, in an attempt to cure a disease such as diabetes, which is treatable.

IMMUNE DEFICIENCIES

The immune system is designed to combat invading microbes, and without it all animals face certain death from a multitude of diseases. The immune system consists of an enormous population of white blood cells that appear in many different forms, the most important of which are the B cells, T cells, and macrophages. B and T cells are lymphocytes that develop in bone marrow and the thymus, respectively. Macrophages are phagocytic blood cells that confront invaders head-on by eating them (a process called phagocytosis), whereas B cells attack them indirectly by producing antibodies. T cells control and coordinate the immune response by releasing signaling molecules called cytokines that recruit macrophages and B cells. T cells also have the remarkable ability to detect invaders that are hiding inside a cell. Even more remarkable, they can force the infected cell to commit suicide in order to control the spread of the infection.

A common form of immune deficiency is called severe combined immunodeficiency-X1 (SCID-X1). This disease represents a group of rare, sometimes fatal, disorders that destroy the immune response. Without special precautions, the patients die during their first year of life. Those who survive are susceptible to repeated bouts of pneumonia, meningitis, and chicken pox.

All forms of SCID are inherited, with as many as half of the cases being linked to the X chromosome (i.e., half of the cases are inherited from the mother). Males born with this disorder, possessing only a single X chromosome, usually die before reaching their reproductive years. SCID-X1 results from a mutation of a gene called *Gamma-c* that codes for the interleukin-2 receptor (IL-2R). There are many kinds of interleukin receptors, also referred to as cytokine

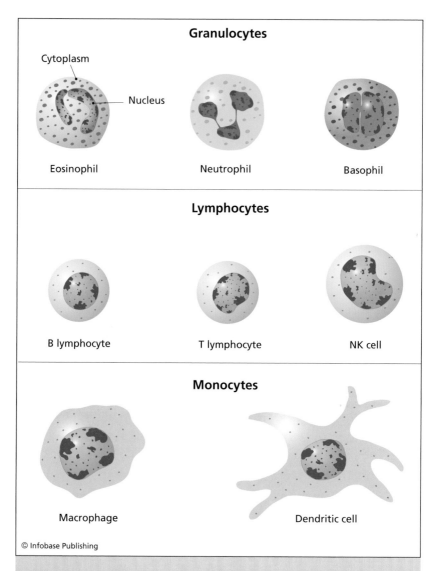

White blood cells. These cells are divided into three major categories: granulocytes, lymphocytes, and monocytes. Granulocytes have a distinctive lobular nucleus, granulated cytoplasm, and all are phagocytic (fat cells, viruses, and debris). Lymphocytes have a smooth morphology with a large round or kidney-shaped nucleus. B lymphocytes are nonphagocytic but produce antibodies. T lymphocytes and natural killer (NK) cells coordinate the immune response and can force infected cells to commit suicide. Monocytes are large phagocytic cells that can engulf bacteria and damaged or infected body cells.

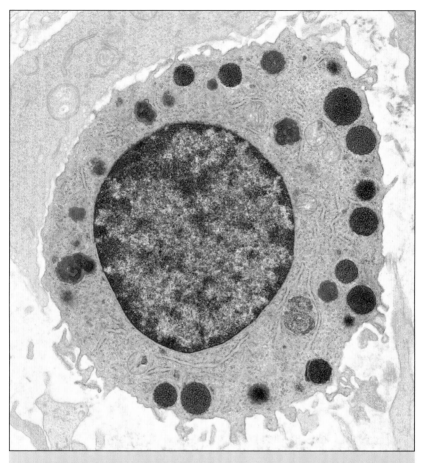

Colored transmission electron micrograph (TEM) of a basophil white blood cell. The nucleus, which stores the cell's genetic information, is green. Basophils are cells of the body's immune system involved in allergic and inflammatory reactions. They secrete the chemicals heparin, histamine, and serotonin, which are stored in granules (red) in the cell's cytoplasm. (*Steve Gschmeissner/Photo Researchers, Inc.*)

receptors, and all of them are crucial for proper communication between the white blood cells. Defective cytokine receptors and the signaling pathways they activate prevent the normal development of T lymphocytes that play a key role in identifying invading agents as well as activating other members of the immune system.

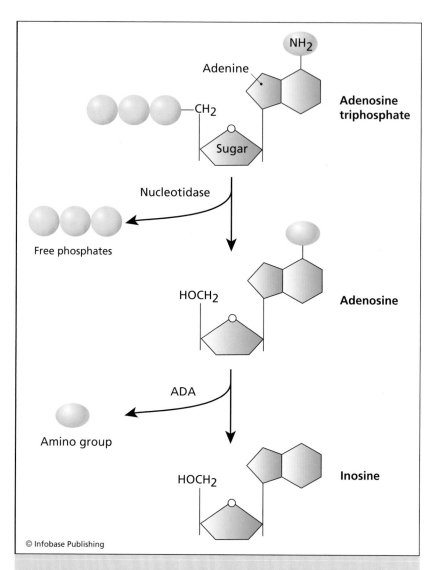

© Infobase Publishing

Recycling adenosine triphosphate. Disassembly begins with the removal of the phosphates by a nucleotidase to produce the nucleoside adenosine, followed by the removal of the amino group by adenosine deaminase to produce inosine. All three components are recycled to make new nucleotides and amino acids.

A second form of SCID is due to a mutation in the *ADA* gene, located on chromosome 20. This gene is active in T lymphocytes, and the gene product, ADA, is required for the recycling of nucleotides. Consequently, a mutation in the *ADA* gene leads to a toxic buildup of adenosine inside the cell, blocking the normal maturation and activity of this crucial member of the immune system. Some patients suffering from ADA deficiency can mount a weak immune response, but in most cases the response is abolished. The conventional treatment, involving a bone marrow transplant, has been successful in saving many lives, but, as with leukemia patients, finding a compatible donor is often impossible.

Gene therapy has been used to treat ADA deficiency and SCID-X1 with moderate success. However, a combination of stem cell and gene therapies was used by Dr. Alain Fischer of the Necker Hospital in Paris to cure SCID-X1, and by Dr. Claudio Bordignon at the San Raffaele Telethon Institute for Gene Therapy in Milan to cure ADA deficiency. Fischer's team extracted bone marrow from affected patients, transfected the hematopoietic stem cells with a healthy copy of the *Gamma-c* gene, and then reimplanted the cells into each patient. Of the eight patients so treated, seven developed a functional immune system. Some of the patients developed a normal T cell count within three months of being treated, which has so far been sustained. Two of the patients subsequently developed a vector-induced leukemia. Bordignon's team, using the same procedure, transfected hematopoietic cells obtained from two patients with a healthy copy of the *ADA* gene. Both patients showed complete recovery and for the past eight years have been able to lead normal lives. To date, 17 out of the 20 enrolled SCID-X1 patients have been successfully treated in London and Paris clinical trials.

LEUKEMIA

Each year, in the United States alone, more than 40,000 adults and 3,000 children develop leukemia, a cancer of the blood cells. There are essentially two kinds of blood cells: red blood cells (RBC) and white blood cells (WBC, or leukocytes). RBCs contain a red protein called hemoglobin and use it to carry oxygen from the lungs to the tissues. White blood cells do not carry oxygen but are part of the body's immune system. Leukemia affects a kind of white blood cell known as a lymphocyte, which spends much of its time in the lymphatic system. Another form of lymphocyte, a myeloid cell, spends its time in the bone marrow or general circulation.

Leukemia affects white blood cells only and can arise in either lymphoid cells (lymphocytic leukemia) or myeloid cells (myelogenous leukemia). The disease has two forms: acute and chronic leukemia. Acute leukemia progresses very quickly and usually destroys the patient's immune system. Chronic leukemia progresses much more slowly, and, even though the leukocytes are transforming, they retain some of their normal functions, so the immune system is not destroyed so quickly or so completely.

The standard treatment for leukemia involves radiation and chemotherapy, which kill the cancerous cells. Extreme forms of this therapy involve the complete destruction of the bone marrow with radiation therapy, after which the patient receives new bone marrow from a suitable donor. The great problem associated with this therapy is the often-impossible task of finding a donor. The best donor is an identical twin, who provides genetically identical, or autologous (or autogeneic), transplant tissue. Non-twin siblings or a parent may also serve as donors, but in these cases, known as allogeneic transplants, rejection of the donated bone marrow is always a threat.

For many patients there simply are no suitable donors, in which case the outlook is grim. Stem cell therapy is now a standard clinical procedure to treat all forms of leukemia with autologous transplants, thus removing the need to find bone marrow donors. Stem cells, isolated from the bone marrow of the affected patient, are induced to differentiate into white blood cell precursors and then grown in culture to increase their numbers. Once these cells

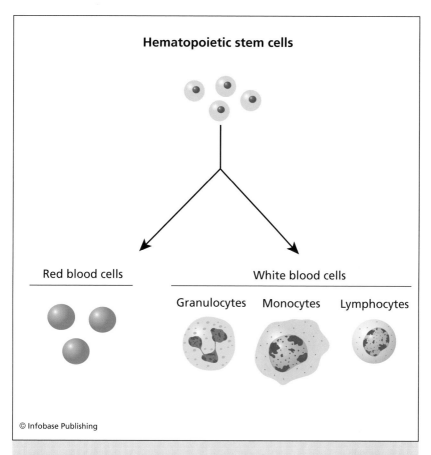

Blood cell types. Blood cells, produced from hematopoietic stem cells located in the bone marrow, are either red or white. Red blood cells have no nucleus and are designed to carry oxygen to the tissues. RBCs get their color from a red oxygen-carrying protein called hemoglobin. White blood cells are nucleated and part of the immune system. They are not involved in oxygen transport and carry no hemoglobin. WBCs are divided into three categories: granulocytes, monocytes, and lymphocytes. Granulocytes and monocytes are phagocytic, whereas the lymphocytes specialize in producing antibodies and in coordinating the immune response.

are collected, the patient's cancerous bone marrow is destroyed and the stem cell–derived blood cells are returned to the patient in order to reconstitute a healthy, cancer-free bone marrow. In 2008, more

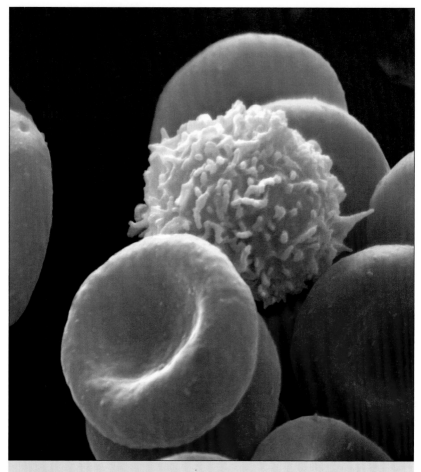

Red blood cells (erythrocytes) and white blood cells (lympho-cytes). (*Eye of Science/Photo Researchers, Inc.*)

than 15,000 American leukemia patients were successfully treated with stem cells.

LIVER DISEASE

Proteins that are eaten for food are broken down to amino acids, which may be used to generate energy or to construct proteins for use. A major by-product in the catabolism of amino acids is ammonia, the stuff of Earth's ancient atmosphere and a molecule that in high concentrations is toxic. Cells deal with the toxicity by converting the ammonia to urea, a much safer molecule that passes out of our bodies as urine. The production of urea depends on the liver enzyme ornithine transcarbamylase (OTC). If OTC is defective, blood levels of ammonia increase rapidly, resulting in coma, brain damage, and death.

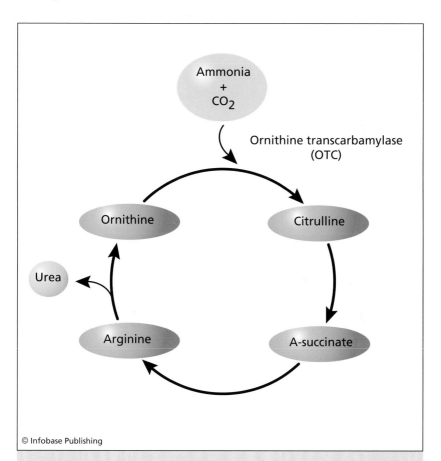

© Infobase Publishing

The urea cycle. Cells in the liver rid the body of toxic ammonia by converting it to urea, which is then excreted by the kidneys as urine. Ammonia and carbon dioxide are added to ornithine to produce citrulline, a reaction that is catalyzed by the enzyme ornithine transcarbamylase. Other enzymes in the cycle produce argininosuccinate (A-succinate) and arginine; the latter is split into urea and ornithine, thus completing the cycle.

A gene therapy trial in 1999 to treat OTC deficiency was terminated abruptly when one of the patients, Jesse Gelsinger, died. Because this disease affects a solid organ, a viral vector carrying a normal copy of the OTC gene was injected into his liver. Jesse Gelsinger's immune response was so extreme that it led to multiorgan failure and death. A safer alternative employs the same strategy used to treat leukemia and SCIDS. Hematopoietic stem cells, isolated from the patient and stimulated to differentiate into liver cells, can be transfected with a vector carrying a good copy of the OTC gene. Reintroduction of these cells into the patient is much safer than injecting naked viruses, since transfected stem cells are much less likely to invoke an immune attack. Once injected into the patient, the stem cells, now partially differentiated, will colonize the liver and produce enough OTC to cure the disease. Preclinical studies on mice have shown that bone marrow–derived stem cells can effectively rescue experimental liver failure and contribute to liver regeneration.

NEUROLOGICAL DISORDERS

There are several neurological disorders that may eventually be treatable with stem cells: These include Alzheimer's disease (AD), Parkinson's disease (PD), Tay-Sachs disease, Huntington's chorea, spinal cord injury, and stroke. Our understanding of Tay-Sachs disease and Huntington's chorea is so incomplete that stem cell therapies for these diseases will not be available for many years. A great deal of progress has been made recently in our understanding of AD, PD, spinal cord injury, and stroke. Thus, it is likely that stem cell therapies for these conditions will be available in the near future.

Alzheimer's Disease

This is a devastating neurological disorder affecting the central nervous system (CNS), which leads to a progressive loss of memory, language, and the ability to recognize friends and family. The average course of the disease, from early symptoms to complete loss of cognition, is 10 years. Alois Alzheimer first described AD in 1907; the disease has since become the fourth-leading cause of death among the elderly. The incidence of AD increases with age and is twice as common in women as it is in men.

The human CNS is divided into the cerebrum (the main portion of the brain including the cerebral cortex), the cerebellum, and the brain stem. The cerebrum is the home of our intellect and the source of our personality. It also processes and analyzes information from all the sensory nerves of the body. A special area of the cerebrum called the hippocampus is important for coordinating memory functions. The cerebellum regulates fine motor control over our muscles, making it possible for us to learn how to play the piano, manipulate fine objects with precise control, and perform other activities that require intricate coordination. The brain stem is in control of our automatic functions, such as the rate at which the heart beats, muscle contraction of the digestive tract, and our respiratory rate. It also controls our ability to sleep and to stay awake.

Alzheimer's begins in the hippocampus. During the early stages, known as preclinical AD, some damage occurs to the brain but not enough to produce outward signs of the disease. Over a period of years, AD spreads to many areas of the cerebrum, leading to the confusion and loss of memory that accompany this disease. Three genes have been identified that are associated with the onset of AD. The first of these is called *Tau,* which codes for a protein needed for the construction of microtubules. The second gene, *App* (amyloid precursor protein), codes for a protein that is embedded in the cell membrane. The third gene, *Sen* (from senilin, also known as presenilin), codes for an enzyme that may be involved in the processing of *App.* Defects in any or all of these genes lead to the extensive death of neurons that is characteristic of AD.

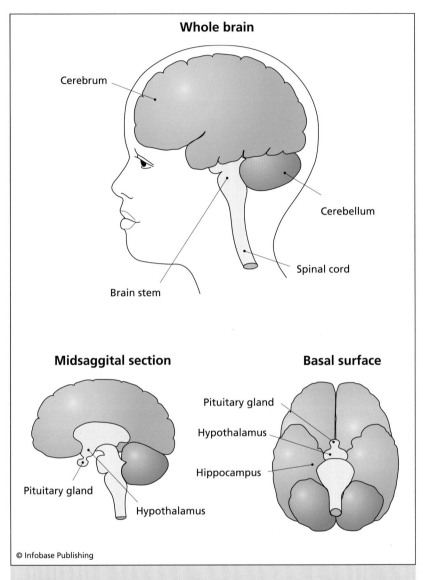

Whole brain

Cerebrum

Cerebellum

Spinal cord

Brain stem

Midsaggital section

Pituitary gland

Hypothalamus

Basal surface

Pituitary gland

Hypothalamus

Hippocampus

© Infobase Publishing

The human central nervous system. The human brain consists of the cerebrum, the cerebellum, and the brain stem, which is continuous with the spinal cord. The brain and spinal cord are called the central nervous system. The pituitary gland, a crucial part of the neuroendocrine system, is connected to the hypothalamus at the base of the brain (mid-sagittal section). The hippocampus, located on the basal surface of the brain, coordinates memory functions.

Stem cells, stimulated to differentiate into neurons and glia cells, may be able to repair the damage to the brain that is caused by AD. Recent experiments with mice and rats have shown that stem cells, injected directly into the brain, can produce functional neurons that make connections with healthy neurons near the lesion. Whether these neurons are making correct connections or not is yet to be determined, and scientists have no idea what effect the growth of these transplanted neurons will have on the psychology of a human patient. The risk of teratoma formation has kept ES cell therapy for AD at the preclinical stage, and this is not likely to change in the near future. But therapies based on AS or iPS cells may be available within the next 10 years.

Parkinson's Disease

This neurological disorder was first described by James Parkinson in 1817 and has since become a serious health problem, with more than half a million North Americans affected at any one time. Most people are more than 50 years old when the disease appears, although it can occur in younger patients. Parkinson's disease is a neurodegenerative disease, affecting neurons in an area of the brain called the substantia nigra, that results in tremors, muscular stiffness, and difficulty with balance and walking.

Until recently, PD was not thought to be heritable, and research was focused on environmental risk factors such as viral infection or neurotoxins. However, a candidate gene for some cases of PD was mapped to chromosome 4, and mutations in this gene have now been linked to several Parkinson's disease families. The product of this gene is a protein called alpha-synuclein, which may also be involved in the development of AD.

Since the neurological damage caused by PD is restricted to one region of the brain, stem cell therapy may be successful in treating this disease. Preclinical research has shown that it is possible to isolate stem cells that can be stimulated to differentiate into

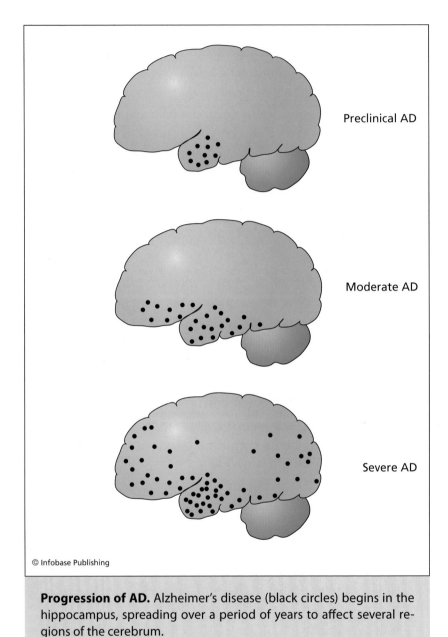

Preclinical AD

Moderate AD

Severe AD

© Infobase Publishing

Progression of AD. Alzheimer's disease (black circles) begins in the hippocampus, spreading over a period of years to affect several regions of the cerebrum.

dopaminergic neurons (i.e., neurons producing a neurotransmitter called dopamine). This is the type of neuron that populates the substantia nigra and is damaged by PD. Injection of these neuronal

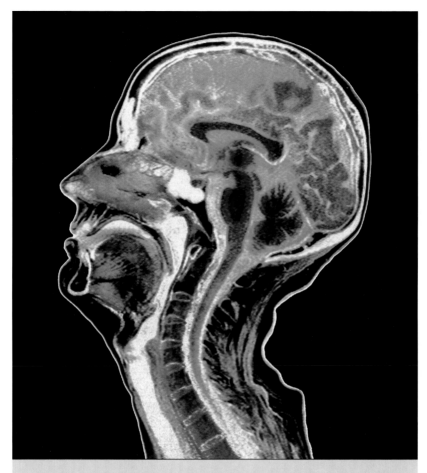

Colored magnetic resonance imaging scan of a sagittal section through the brain of a 51-year-old male, showing cerebral atrophy. Atrophy of parts of the cerebrum of the brain occurs in various disorders, including stroke, Alzheimer's disease, and AIDS dementia. The green/red brain is "sliced" through, showing the large folded cerebrum (at top). Here the area of the upper cerebrum affected by atrophy is colored dark red. (*Simon Fraser/Photo Researchers, Inc.*)

stem cells into the brains of affected mice with Parkinson's relieved some of the disease symptoms, particularly the loss of motor control that is characteristic of this disease. In another study, Steven Goldman and his team at Cornell University Medical College in New York produced a homogeneous population of dopamine-producing neurons from human ES cells by growing them in the presence of glia cells, whose normal role is to support and maintain the growth of neurons. When the new dopamine neurons were transplanted into the brains of rats with the symptoms of Parkinson's disease, the animals recovered almost entirely. The researchers observed, however, that the transplanted cell population was not perfectly homogeneous but contained some undifferentiated stem cells that could have formed teratomas. Although not confirmed (the rats in the study were killed before any such tumors developed), it was serious enough to keep this study from progressing to a clinical trial.

As previously mentioned, Rudolf Jaenisch and his team at the Whitehead Institute for Biomedical Research in Cambridge, Massachusetts, produced iPS cells from rats suffering from Parkinson's disease. The cells were differentiated into neural precursor cells that went on to form neurons and glia cells after being injected into the brains of fetal mice. The neural precursor cells were differentiated into dopamine neurons and then injected into the brains of adult rats suffering from Parkinson's disease. After a time, the researchers noticed a marked improvement in the rats' motor skills. In 2009, they repeated the first part of this experiment on four patients suffering from PD and were able to produce patient-specific dopaminergic neurons. These very promising results will soon be tested in a clinical trial, at which point the dopaminergic neurons will be injected into the brains of the PD patients.

Spinal Cord Injury

Damage to the spinal cord, caused by automobile accidents or falls from high places, can make it impossible for the brain to control the extremities and internal organs, such as the heart and lungs. The severity of the damage depends on how close to the brain the spinal cord injury is. If the individual's neck is broken, he or she may end up being a quadriplegic (a person unable to move the arms and legs) and may not be able to breathe properly. If the damage to the spinal cord is near the middle of the back, the patient will be paraplegic (unable to move the legs) but will retain control over the arms and lungs.

Repairing a damaged spinal cord is extremely difficult for two reasons. First, the neurons that were destroyed must be replaced, and the replacement neurons must make the proper connections to bridge the damaged area. Second, once the new neurons are in place, they must be insulated, much as an electrical wire is insulated, before they can work properly.

Neurons are remarkable cells, specially designed for communication and the construction of elaborate circuits. Signals enter a neuron through its dendrites and leave by its axon (some neurons may have more than one axon). The neural network, or circuitry, is established when the axon from one neuron makes a connection with a dendrite of another neuron. Neural circuitry, particularly in the spinal cord, does not work well unless the axons are insulated with a myelin sheath. The myelin is constructed from cells called oligodendrocytes and Schwann cells that wrap around axons to form a protective multilayered sheath. The oligodendrocytes, located in the central nervous system, can myelinate more than one axon at a time. The Schwann cells, located in the peripheral nervous system (i.e., any nerves outside the CNS, such as those in the toes and fingers), wrap around a single axon.

Stem cell therapists attempting to repair a damaged spinal cord must provide neurons to reestablish the circuit and oligodendrocytes for insulation. Dr. Ronald McKay, a stem cell researcher at the NIH, and Dr. Mirodrag Stojkovic, at the Cellular Reprogramming Laboratory in Valencia, Spain, have shown that mouse ES and rat AS cells can repair some neural damage when injected into rats. But whether stem cells will be able to reestablish normal circuitry and, at the same time, re-myelinate the axons is yet to be determined. In Stojkovic's study, the treated animals showed partial recovery of motor activity one week after injury. Thus, it would seem likely that

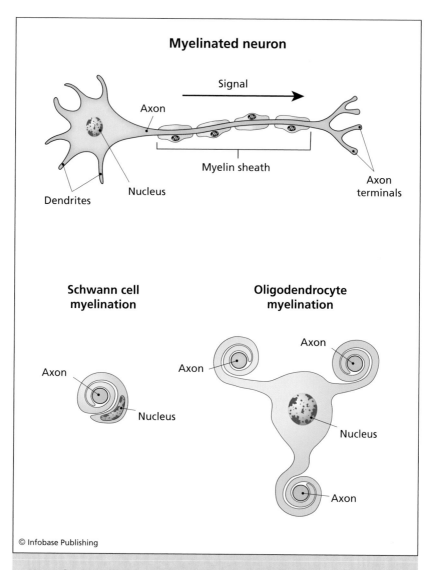

Neural signaling and myelination. Communication between neurons is much more efficient when the axons are insulated with a myelin sheath. Myelin is made by Schwann cells or oligodendrocytes that wrap around the axon. Oligodendrocytes can insulate more than one axon at a time.

some re-myelination had occurred. Moreover, since they used AS cells, it would be possible to treat humans in this way without the worry of teratoma formation or immune rejection.

Stroke

A stroke occurs when the blood supply to the brain is interrupted, leading to the death of brain cells. As a result, the affected part of the brain is unable to function, leading to partial or complete paralysis of various parts of the body. A stroke can lead to permanent neurological damage and is often fatal. Blood flow to the brain may be disrupted for several reasons: narrowing of the small arteries in the brain, hardening of the arteries leading to the brain, or a blood clot. (When a clot breaks away and travels to another part of the body, it is said to have embolized.)

Stroke is a very common problem and is currently the leading cause of adult disability in the United States and Europe. Nearly 800,000 Americans suffer a stroke each year, and of those more than 140,000 are fatal. In 2009, Americans will pay about $70 billion for stroke-related medical costs. Dominant risk factors include age, diabetes, high blood pressure, high cholesterol, and cigarette smoking.

Dr. James Fallon and his associates at the University of California, Irvine, have shown that it is possible to activate the body's own stem cells to repair neural damage caused by a stroke. In 2009, his team was able to activate AS cells by injecting transforming factor α (TF-α) into a stroke-damaged area of a rat's brain. The activated stem cells not only repaired the neural damage, but the rats showed a near complete (99 percent) behavioral recovery. Fallon's team also found that the response is very specific, in that an injection of TF-α into a normal rat does not lead to a stimulation of stem cells. Damaged areas of the brain seem to prime the stem cells for repair, making them responsive to the influence of growth factors. In addition to solving the problem of immune rejection (when ES cells are used), this procedure also removes the technical difficulties associated with in vitro differentiation and with injecting stem cells into the body.

The use of growth factors does not, however, resolve the problem of memory retention, nor does it ensure the maintenance of a human patient's psychology. Moreover, growth factors are dangerous drugs to administer because there is a risk they might overstimulate some cells, forcing them to divide inappropriately and setting the stage for cancer development. Nevertheless, this is a very promising approach that may soon be tested in clinical trials.

ORGAN FACTORIES

Scientists are interested in using stem cells to grow whole organs in the laboratory that could be used to replace a defective heart, kidney, pancreas, lung, or liver. As discussed in chapter 7, a research team at Harvard Medical School has grown a miniature cow's kidney from cultured ES cells. The ES cells were obtained by cloning the cow and allowing the fetus to develop up to the stage where kidneys normally begin forming. By isolating cells from the fetus, in the area where the kidneys normally form, the researchers, in effect, isolated an embryonic kidney, which continued growing in cell culture. This cultured kidney was able to produce urine when implanted back into the cow from which the embryo was cloned.

This research demonstrates the feasibility of using stem cells to produce organs, and if successful its results will go a long way toward relieving the chronic shortage of organ supplies for transplant surgery. Of equal importance is the effect it will have on relieving the ethical and social problems associated with the current practice of obtaining organ donations from family members, who must be subjected to severe surgery even though they may be perfectly healthy themselves and from which they may not recover. Unfortunately, the progress of this line of stem cell research has not been encouraging. Indeed, all of the companies that began this type of work in 2001 have since abandoned the effort. This research, conducted primarily by biotech firms, is extremely difficult and would likely take more than 20 years before the company could produce a functional organ. By that time, simpler therapies, which could repair damaged organs without the need for a transplant, may have been developed.

FUTURE PROSPECTS

By far the most successful application of stem cells is for the treatment of leukemias and lymphomas. The needed stem cells are easy to obtain, they do not cause GVHD or form teratomas. Moreover, these diseases are relatively simple, and the transplanted stem cells do not have to reconstruct a tissue or an organ with a complex structure. These facts are relevant to diabetes and liver disease as well and should facilitate the development of stem cell therapies for these diseases.

Stem cell therapies for the heart and the nervous system require a precise integration of the transplanted cells into complex organ systems. This is bad enough, but other problems such as the threat of GVHD and the formation of teratomas have blocked the progression of this research to clinical trials. The use of iPS cells could solve the problem of GVHD, but there is still much research needed before neuron progenitors, derived from iPS cells, could be used in a clinical trial. All stem cell preparations will have to be rigorously screened to ensure the removal of all undifferentiated cells. Scientists at Geron Corporation have produced oligodendrocyte progenitor cells from human ES cells, which they hoped could be used to treat spinal cord trauma. Early in 2008, they announced that the cells were about to be tested in a Phase I clinical trial, but just weeks before the trial was to begin their permission to proceed was revoked by the FDA over concerns that the cells were not fully characterized and could thus form teratomas. In 2009, they obtained permission to test their cells in a Phase I trial, but only after convincing the FDA that their cultures were free of undifferentiated ES cells.

Commercialization of Human Stem Cells

Conducting biomedical research is very expensive. Before a drug or therapy can be licensed for use as a standard medical treatment, it has to pass through a review gauntlet consisting of preclinical research using laboratory animals, followed by four stages of clinical trials in which the new treatment is tested on human volunteers. According to the Tufts Center for the Study of Drug Development, the average cost is $802 million, but can be as high as $3 billion and can take up to eight years. Governmental granting agencies, such as the National Institutes of Health (NIH) in the United States, allocate billions of dollars each year for basic biomedical research, but it is never enough to cover clinical research. That is why clinical trials involving stem cell therapies and many others, including gene therapies, are backed financially by pharmaceutical companies. Indeed, without the backing of wealthy companies, many drugs and medical therapies would never be brought to market or used to treat the general public. Instead, they would remain within the confines of research laboratories, useful only for curing mice and rats.

THE STOCK MARKET

Where do pharmaceutical companies get the money to fund clinical research? Much of it comes from profits earned by selling their products, but a great deal of it comes from selling shares in their companies to the general public. The buying and selling of company shares, also referred to as stocks or securities, are handled by a stock exchange like the New York Stock Exchange (NYSE) or the National Association of Securities Dealers Automated Quotations (NASDAQ), both of which are based in the United States. Most of the pharmaceutical and biotechnology companies are listed on the NASDAQ stock exchange. (NASDAQ gets its rather awkward name from the fact that buying and selling of stocks, including stock price quotations, is completely automated through the use of computers and the Internet. The NASDAQ has no trading floor, unlike the NYSE, where stockbrokers handle the transactions directly.) In 2009, the NASDAQ listed more than 250 biotech and pharmaceutical companies, of which only five or 10 are actively involved in stem cell research. (Most of the preclinical work is undertaken in academic research institutes.) The annual value of all biotech and pharmaceutical companies (referred to as market capitalization) is more than $300 billion.

The stock market may seem like a long way from a research laboratory, but the ability of pharmaceutical companies to fund clinical research depends on how successful they are at selling their stock and attracting investors. Not surprisingly, their ability to do so depends on how good their product is, and for biomedical research, the best yardstick is performance of a drug or therapy in a clinical trial. If a company advertises a drug that was proven effective in preclinical trials, many investors will want to buy shares in the company, and the more investors who buy, the more valuable the stock becomes. Investor confidence can drive the stock price from $1/share to $100/share, practically overnight. This is the hope of most investors in the pharmaceutical sector—that a drug will be proven effective after passing through all the clinical trials and will go on to earn the company and the shareholders a great deal of money.

This happy scenario played itself out many times during the early 1990s, when both the economy and investor confidence were high. Many biotech and pharmaceutical companies saw the value of their stocks increase from a few dollars/share to nearly $100/share. Investors gave large sums of money to these companies whether they had a strong product or not. Biotech companies seemed like a sure bet, and if their product succeeded the companies could earn billions of dollars. By the late 1990s, this overly optimistic outlook had all but evaporated. Biotech companies that were trading at more than $60/share dropped to $10/share in less than a year. By 2009, some of these companies were trading at less than $5/share.

The financial trouble experienced by the biotech and pharmaceutical companies was caused in part by a general decline in economies all over the world the ever-present threat of war, the escalation of terrorists' attacks against the Western world, including the destruction of the New York World Trade Center in 2001, and the global recession that began in 2007. Investor confidence in the stock market as a whole was also badly shaken by a number of accounting frauds that attempted to make companies' products look better than they really were. One such case affecting the biotech sector involved a company called ImClone Systems, which claimed to have developed a very effective drug called Erbitux to fight cancer. Government investigators discovered, however, that the claims were exaggerated, and the company's chief executive officer was tried and convicted of securities fraud. Before the scandal was exposed in early 2002, ImClone's stock sold for $80/share, dropping to less than $10/share by the end of the year. The company is now a subsidiary of Eli Lilly & Company.

Honest biotech companies, and there are many of them, have to contend with these problems, but fraudulent behavior is only a small part of the problem. Many biotech companies have suffered financially simply because they could not bring their product to market or they were not progressing as fast as investors would like. A medical therapy or a new drug may have looked good initially but failed in a clinical trial, or the results of a clinical trial may suggest that further preliminary work is required before the therapy can move to the next stage. Investors usually are not scientists and do not understand the technical difficulties associated with the biotech products. Producing a reliable drug or stem cell therapy rarely happens in a time frame that keeps investors happy.

AASTROM BIOSCIENCES

Aastrom is located in Ann Arbor, Michigan, and was formed in 1991, going public in 1997. This company is primarily involved in the isolation of adult stem (AS) cells from bone marrow and umbilical cord blood. The company's Tissue Repair Cell (TRC) technology involves the use of the patient's own bone marrow as a source of the AS cells used to treat chronic diseases affecting the heart, bone, vascular system, and neural tissue. The company also provides a kit for isolating stem cells, called the AstromReplicell System, which provides standardized methods and reagents for directed differentiation.

Currently, the company is using cardiac repair cells (CRCs), produced with its TRC technology, in a Phase II clinical trial for the treatment of dilated cardiomyopathy. In addition, vascular repair cells (VRCs) are being tested as a treatment for critical limb ischemia (i.e., the VRCs will be used to regenerate damaged blood vessels) and bone repair cells (BRCs) have entered a Phase III trial for the treatment of osteonecrosis of the femoral head. The FDA is close to approving the BRCs as a medical treatment. The company's neural repair cells are at the preclinical stage of development.

Aastrom's stock value reached a peak of $8.50/share in 2000 but dropped below $1/share by 2009. With more than 46 million shares originally sold, this company has lost more than $300 million since it began selling shares. Much of the decline has been due to the difficulties involved in obtaining a clear-cut therapeutic effect of its cells in clinical trials. Success in 2007 with the Phase I clinical trial to treat cardiomyopathy and the launch of a Phase II study in 2008 are expected to improve the company's future.

STEMCELLS, INC.

This company distinguishes itself from Aastrom Biosciences by focusing its attention on cells that may be used to treat disorders of the central nervous system (CNS), such as Parkinson's disease and Alzheimer's disease. StemCells, Inc., has discovered and characterized a human CNS stem cell line (HuCNS-SC) and has perfected the transplant procedures. It has also isolated insulin-producing stem cells to treat diabetes and other stem cells for treating diseases of the liver.

The company launched its first clinical trial in May 2006 to test the effectiveness of the HuCNS stem cells in the treatment of Batten disease, a neurological disorder characterized by a buildup of lipofuscin in the brain that leads to cell death. The objective in this Phase I trial is to evaluate the safety of the stem cells. The trial, including a lengthy follow-up, is expected to run for five years. The HuCNS stem cells may also be effective in the treatment of Alzheimer's disease and other neurological disorders, but the work is in the preclinical stage.

StemCells, Inc., located in Palo Alto, California, was formed in 1988 as Cytotherapeutics, changing to its current name in May 2000. A share in this company reached a maximum of $11 in 2001, dropping below $2 by 2009. The company has sold more than 3 million shares but lost more than $59 million since 2000. A successful outcome to their clinical trial may be crucial to the survival of this company.

GERON CORPORATION

In winter 2000, stock prices for Geron Corporation, located in Menlo Park, California, were as high as $70/share, but share prices dropped precipitously to $10/share at the beginning of 2002, falling below $5/share by 2009. This near-collapse of the company's stock value amounted to many millions of dollars lost, but despite this Geron remains the strongest and wealthiest company involved in stem cell research.

In 1996, just months after Ian Wilmut and his team at the Roslin Institute in Scotland announced the cloning of Dolly, the Scottish government decided to withdraw further funding for the project, possibly because they felt the research was too controversial. The institute managed to obtain alternative funding from two biotech companies: One was PPL Therapeutics, based in Scotland, and the other was Geron Corporation. Geron's involvement in the work at Roslin was so complete that in 1999 the corporation bought Roslin Bio-Med, a company formed by the Roslin Institute, and along with it the patent for the nuclear transfer procedure developed by Drs. Wilmut and Keith Campbell to clone Dolly. In addition to the original patent, Geron now has more than 60 patent applications pending for nuclear transfer technology.

Geron's involvement in stem cell research began in 1998, when it funded the work of Dr. James Thomson at the University of Wisconsin (UW), who was among the first to isolate and characterize embryonic stem cells (ES) cells from humans. In 2002, UW and Geron worked out an agreement in which Geron received exclusive rights to the ES cells when they are used to treat heart disease, diabetes, and neurological disorders, while UW obtained exclusive rights to the cells for treatments involving bone, blood, and liver ailments. Consequently, researchers wishing to use those cells to treat heart disease must pay a licensing fee to Geron, whereas researchers wishing to use the ES cells to treat leukemia must pay UW a fee.

Owning patents for nuclear transfer technology and for many ES cell lines places Geron in an extremely powerful and potentially lucrative position. The application of stem cell therapy often includes the use of animal cloning procedures. For example, a privately owned biotech company called Advanced Cell Technology is planning to produce ES cells by cloning human eggs, rather than fertilizing them in vitro as was done for all of the current ES cell lines. By doing this, they avoid paying Geron licensing fees for their cells, but they still have to pay a fee for using Geron's nuclear transfer procedures.

Research at Geron, in collaboration with scientists at the University of California, Irvine, is now focused on the use of ES cell–derived neural cells to treat nervous system disorders, such as Alzheimer's disease and spinal cord trauma. To that end, oligodendrocyte progenitor cells have been isolated from human ES cells grown in culture. These cells were shown to improve hind-leg movement and to partially restore functional locomotor behavior in a rat model of spinal cord injury. The company has extended these studies to humans in a Phase I clinical trial that began in 2009.

Many view the commercialization of stem cell research, especially when it involves human ES cells, as a mixed blessing. The recent controversy involving a painkiller called Vioxx (rofecoxib), made by Merck, one of the largest pharmaceutical companies in the world, is a case in point. Merck aggressively promoted Vioxx even though some researchers had concerns that it could cause heart failure. A class-action lawsuit was launched against the company in 2006 after several patients died while taking the drug. Thousands of documents relating to Vioxx were made public as part of the legal action. A review of these documents in 2008 showed that Merck was disguising its role in research that showed Vioxx to be effective and harmless. This was accomplished by omitting the names, or downgrading the involvement, of drug-industry writers and adding the names of academics (i.e., hiring ghostwriters) who were not involved in the research. In effect, Merck would conduct a clinical trial, having its employees design the trial, analyze the data, write the paper, and then toward the very end recruit academic authors to put their names on the paper to give it a seal of supposed authenticity. The analysis of the Merck documents was conducted by Dr. Joseph Ross and his associates at the Mount Sinai School of Medicine in New York and was published in the *Journal of the American Medical Association* in 2008.

Without the money large corporations bring to medical research, many potentially powerful therapies would never see the light of day. But the Vioxx controversy has made it clear that the mixture of science and profit does not always smell so sweet, leading as it sometimes does to exaggerated claims, secretive results, ethical conflicts, and fraud. No one would suggest that corporations should not be involved in this kind of research, but there are many problems to resolve before the association benefits everyone.

Reality Check

Stem cells are demonstrating great potential for forming the basis of medical therapies that could cure cancer, neurological diseases, and spinal cord injuries. But in their enthusiasm for these therapies, many scientists have given the general public an unrealistic impression of what can actually be achieved. In theory, of course, the sky is always the limit, but scientists interested in using stem cells to repair damaged organs fail to make it clear that there are serious obstacles to overcome before these therapies become a reality.

The versatility of stem cells and the promise they hold is without question, but devising a practical method to realize that promise will be extremely difficult. Injecting stem cells into a patient leads to the same problem that transplant surgeons have been trying to overcome since the 1950s: namely, the rejection of those foreign cells by the patient's immune system. Even if scientists can overcome this problem, they are left with an even greater challenge in trying to ensure the cells will go where they are supposed to go and do what they are supposed to do without forming a cancerous tumor, thereby damaging the very organs they are supposed to repair. This is particularly worrisome in the case of a brain disorder, where stem cells are expected to make extremely complex repairs while leaving the patient's original memories and psychology intact. This would be a tall order even if scientists had a clear understanding of memory and personality functions, but at the present time they do not.

Other researchers have claimed that they will be able to grow vital organs, such as a heart or kidney, by culturing stem cells in special ways. But acquiring the knowledge to make this possible will be so difficult and will take so long that the idea occupies the realm of science fiction. This chapter will explore the many claims that have been made by stem cell researchers and will take a close, critical look at the proposed applications while emphasizing hard realities that are likely to keep most of these therapies on the drawing board for a very long time.

GRAFT-VERSUS-HOST DISEASE

Stem cell therapies all require that the cells, usually isolated from human embryos, be injected into a patient, where they will effect whatever repairs are necessary. This is equivalent to a tissue or an organ transplant, and, like any conventional transplant, the stem cells will be attacked by the patient's immune system leading to a condition known as graft-versus-host disease (GVHD).

Transplants are categorized according to the source of the transplant tissue. Autografts are transplants of tissue from one part of an individual to some other area of the same individual. A common example involves bypass surgery in which an artery from the leg is transplanted to the heart in order to repair or bypass a blocked coronary artery. Because the artery was obtained from the patient there is no threat of immune rejection. A second example involves bone marrow transplants, commonly used to treat leukemia. It is sometimes possible to obtain donor bone marrow from an identical twin, in which case the cells are genetically identical to the eventual host, and, again, there is no threat of immune rejection. An autograft is also known as an autogeneic transplant. A syngeneic transplant, involving identical twins, is synonymous with an autograft. Allografts involve tissue that is taken from one individual and transplanted into another, unrelated, individual. This type of transplant will be attacked by the immune system because it is genetically dissimilar to the host. The great majority of transplants involving organs or tissues are allografts. The proposed transplantation of human embryonic stem (ES) cells into patients suffering from Alzheimer's disease is another example of an allograft. An allograft is also called an allogeneic transplant. Xenografts are transplants between different species. Transplanting the heart of a pig into a human is an example of a xenograft; so too is the transplantation of human ES cells into experimental mice or rats.

Autografts do not invoke an immune reaction, but transplant operations involving allografts or xenografts lead to GVHD. The current treatment for GVHD is to give the patient drugs that inhibit the immune response. All allograft transplant patients must take these drugs, called immunosuppressants, for the remainder of their lives. (This will also have to be done for xenografts, if and when they are performed on humans.) Immunosuppressants act by blocking the activity of white blood cells, called T lymphocytes, that are responsible for hunting down and destroying invading microbes and infected cells. The immune system cannot distinguish between a deadly microbe and transplanted tissue; both are attacked and, if possible, destroyed.

A considerable amount of survival data, both for the transplanted organ and the patient, has been collected since organ and tissue transplants were first attempted in the 1950s. Prior to the introduction of immunosuppressants in the 1980s, transplant patients

SURVIVAL OF TRANSPLANT PATIENTS AND TRANSPLANTED ORGANS IN THE UNITED STATES

ORGAN	1 YEAR PATIENTS	1 YEAR ORGANS	5 YEAR PATIENTS	5 YEAR ORGANS
Kidney	94.4	89.0	81.9	67.5
Heart	87.8	87.3	74.4	73.2
Liver	86.3	82.0	72.0	67.6
Lung	83.3	82.5	47.3	46.1
Pancreas	96.7	80.1	88.1	50.6

Note: Table values are percentages. The data was compiled from information provided by the Organ Procurement and Transplantation Network for 2008.

rarely survived for more than a few weeks after the operation. After 1980, when two powerful immunosuppressants, called cyclosporine and tacrolimus, were introduced, the one-year survival went up dramatically, but long-term survival is still poor owing to a slow chronic destruction of the transplant by the immune system. Thus, even with a daily dose of immunosuppressants, the transplanted organ is likely to fail.

Several reports have appeared, both in the scientific literature and in the news, concerning the use of ES cells to treat neurological disorders. In one such study, mouse ES cells were injected into rats suffering from a surgically induced neurological disorder. The stem cells not only migrated to the appropriate area but repaired much of the damage. However, what is not generally reported in the news is the fact that the animals receiving the stem cells are on powerful immunosuppressants or are so-called knockout rats whose immune systems have been inactivated through genetic engineering.

Chemically or genetically suppressing the immune system means, of course, that these animals are especially susceptible to disease and must live out their lives in a sterile environment. These same problems would affect a human patient subjected to this kind of therapy, and yet scientists have not made this point clear, leaving the public with the impression that therapies based on ES cells could be available in the very near future. Clearly, the promise of ES cell therapy has been greatly exaggerated. Hope remains, however, because there are four alternatives to the use of ES cells: adult stem (AS) cells, umbilical cord stem (UCS) cells, growth factors to stimulate and mobilize the body's own stem cells, and finally induced pluripotent stem (iPS) cells.

SPONTANEOUS IN VITRO DIFFERENTIATION

A major problem associated with the culturing of stem cells is uncontrolled differentiation. Stem cells in culture for six months to a year begin to differentiate into fibroblasts, or cells that look very much like fibroblasts. Once this happens, the researcher has to start a fresh culture. This is easy in the case of AS cells, because the source is not ethically contentious, but is becoming increasingly difficult in the case of human ES cells, where starting a new culture means destroying more human embryos.

Spontaneous differentiation is one of the reasons scientists working with human ES cells are interested in having free and ready access to human embryos. It has not been possible to resolve this very difficult problem, nor is it likely to be resolved in the near future. In the meantime, as the available cultures of ES cells differentiate, investigators anxious to produce medical therapies based on ES cells are finding it increasingly difficult to continue with their work. Fully differentiated ES cells cannot be used for therapies. Moreover, funding of stem cell research by agencies such as the National Institutes of Health (NIH) is based on merit and the perception that the research will lead to a practical therapy, something that is difficult to achieve when researchers have an insufficient number of cultures to study.

The interest in ES cells is also driven by the race to the patent office. This is an important element in a field where pharmaceutical and biotech companies stand to earn a great deal of money should they succeed in producing a workable therapy. It is also an inevitable outcome when, as described in chapter 6, expensive research is funded by stockholders. It is this arrangement of research and economic priorities that is behind the urgent calls for freer access to human ES cells. In the minds of many scientists and the heads of biotech companies, ES cells represent the quickest way to medical therapies, international fame, and the front door of the patent office. One only needs to look at the intense media coverage of the team that cloned Dolly the sheep to see this dynamic in action. ES cell researchers may achieve the therapies they seek, but the legal environment, the production of iPS cells, and the work of Verfaillie and her team (see chapter 2) have already shifted the emphasis away from human ES cells.

WANDERING STEM CELLS

Conventional stem cell therapy involves injecting the cells into the patient and letting them home in on the target tissue, and there is good evidence available to suggest that these cells, after partial in vitro differentiation, do just that. For example, stem cells that have been stimulated in vitro to become hematopoietic cells will migrate to the bone marrow when injected into the body. However, the number of studies is still too small to guarantee that accurate targeting will always occur. Misdirected stem cells could be a very serious problem for therapies aimed at spinal cord disorders or diseases of the brain. There is currently no method for directing a stem cell to differentiate into a brain-specific neuron, as opposed to a spinal cord–specific neuron or neurons of the peripheral nervous system (such as nerves in the heart, gut, or hands). To improve targeting, the stem cells can be injected directly into the brain or spinal cord, but there is no guarantee that all of them will stay there. They may migrate to other neural tissue. Thus, an attempt to treat a brain condition such as Parkinson's or Alzheimer's may result in serious complications if therapeutic stem cells colonize the spinal cord or the peripheral nervous system. It is generally considered a wise policy to avoid trying to fix things that are not broken, but stem cells are programmed for making repairs, and that is what they will try to do wherever they land. In the current example, this could lead to severe disruptions in the normal functioning of the spinal cord. Conversely, a therapy to treat spinal cord trauma could result in severe brain damage.

STEM CELLS AND CANCER

Stem cells have one thing in common with cancer cells—the ability to proliferate indefinitely. This similarity has worried scientists and physicians for some time because it could mean that the use of stem cells to treat a disease may result in the seeding of cancers throughout the body.

With this in mind, researchers at the National Institute of Neurological Disorders in Bethesda, Maryland, began looking for molecular similarities between these two kinds of cell. They found that cancer cells and ES cells both express a protein called nucleostemin. The exact function of this protein is not yet known, but it appears to be a molecular switch that controls cell division. If it turns out that cancer cells escape the normal inhibitions of the cell cycle by activating nucleostemin, this would imply that stem cells have a similar ability and will have to be handled with extreme caution. The use of partially differentiated stem cells may reduce the risk of cancer formation, but additional studies will be needed to reveal more about the body's own stem cells and the controls that keep them in check.

GROWING ORGANS FROM STEM CELLS

In summer 2002, researchers at the biotech company Advanced Cell Technology published a paper in the journal *Nature Biotechnology* in which they described their attempts to grow kidneys from cloned

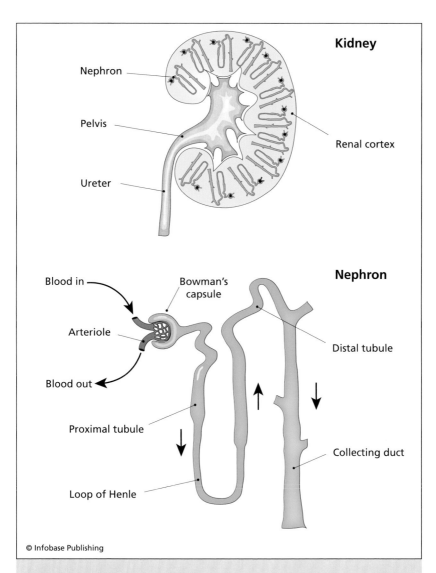

Kidney

Nephron

Pelvis

Ureter

Renal cortex

Nephron

Blood in

Bowman's capsule

Arteriole

Distal tubule

Blood out

Proximal tubule

Collecting duct

Loop of Henle

© Infobase Publishing

Anatomy of the human kidney. A cross section of a kidney is shown at top, and the functional unit, the nephron, is shown enlarged at the bottom. Each kidney contains millions of nephrons arranged throughout the cortex (the size of the nephrons is exaggerated for clarity). The collecting ducts drain into the pelvis, and the urine is carried to the bladder by the ureter. The direction of urine flow through the nephron is indicated by the straight arrows.

cow embryos. Entitled "Generation of histocompatible tissues using nuclear transplantation," the article was also intended to demonstrate the effectiveness of therapeutic cloning, whereby an animal is cloned and the subsequent embryos are used as a source of histocompatible (immune-system compatible) stem cells. Although the company has since abandoned this line of research, the problems they had are still relevant today and show clearly just how difficult such an undertaking can be.

Real kidneys have a very complex structure that poses a huge challenge to anyone trying to grow them in culture. The human kidney, about the size of a child's fist, filters the blood to remove waste products, such as urea and excess salts and other ions that build up as a normal by-product of metabolism. The functional unit of a kidney is an intricately constructed tubule called the nephron, which consists of several distinct regions: Bowman's capsule, the proximal tubule, the loop of Henle, and the distal tubule. Blood plasma (the fluid portion minus the blood cells) enters the nephron at Bowman's capsule, where the incoming arteriole branches out into a capillary bed (the blood cells are too large to enter the nephron). The combination of Bowman's capsule and the capillary bed is known as the glomerulus.

The blood pressure, driven by the heart, forces the plasma out of the capillaries and into the proximal tubule, after which the plasma travels the whole length of the tubule and is processed along the way. The basic strategy of the nephron is to reclaim, by tubule reabsorption, compounds and ions the body needs, while allowing the waste material, such as urea, to pass completely through to the collecting ducts and, finally, out of the body by way of the ureter and bladder. Aiding in this remarkable process is a diverse population of cells that are distinct for each region of the tubule.

The attempt to grow a kidney in culture was headed by Dr. Robert Lanza, an American stem cell researcher. His team began by cloning cows from ear epithelium and then allowing the embryo to

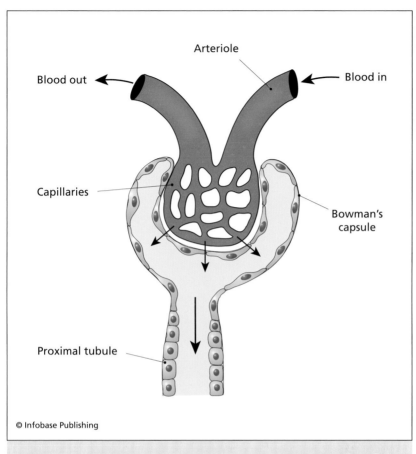

Bowman's capsule. Blood, under pressure from the beating heart, enters the capillaries. The blood plasma (i.e., fluid minus blood cells) diffuses across the capillary membranes, then enters the nephron by diffusing across the single cell layer of Bowman's capsule (small arrows). The plasma passes through the whole length of the nephron, beginning in the proximal tubule (large arrow).

reach an early fetal stage of development (three to four months old), by which time the researchers were able to recognize and isolate embryonic kidney cells. These cells were cultured on a collagen-coated renal unit to give the artificial organs, known as embryoid kidneys,

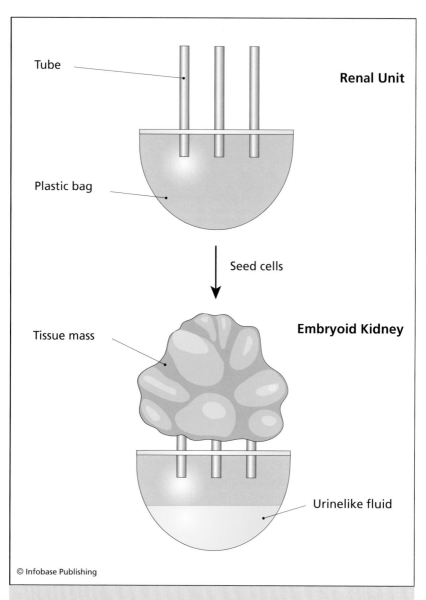

Tube

Renal Unit

Plastic bag

Seed cells

Tissue mass

Embryoid Kidney

Urinelike fluid

© Infobase Publishing

Embryoid kidney. A renal unit (top) is constructed from collagen-coated tubes and a plastic bag. The unit is seeded with embryonic kidney cells, which form a tissue mass containing loosely organized nephronlike structures. Fluid excreted by the embryoid kidney is collected in the plastic bag.

their final shape and to provide a convenient way to collect the fluid excreted by the cells. The embryoid kidneys were implanted under the skin of the adult cow from which the clones were derived (to avoid immune rejection) where they continued producing a urine-like substance.

Although the science community greeted it with some enthusiasm, there are many problems associated with this study. First, the investigators used embryonic kidney cells, not stem cells. Thus, it is not surprising that these cells, once isolated, tried to construct a kidney. The study fails to answer the question of whether stem cells can be used to grow artificial organs. Second, despite being cultivated from kidney cells, the final organ bears no resemblance to a mammalian kidney. Although the cells formed nephronlike structures, these structures have a very crude anatomy, lacking regional variations such as the proximal tubule, loop of Henle, and the distal tubule. The embryoid cells failed to construct a kidney because once they were removed from the body and cultured in vitro they lost the developmental cues that are crucial for normal organogenesis. As a consequence, embryonic cells may form some of the parts, such as a nephron, but they do not know how to organize those parts into a truly functional, anatomically normal organ. Third, a crucial component in this process is the precise interactions between the developing nephron and the arterioles that must form a capillary bed inside every Bowman's capsule. Without the necessary instructions, cultured embryonic cells form an embryoid body, an assemblage of cells that looks more like a tumor than an organ. Lanza's team presented no evidence to document vascularization of the embryoid kidney after implantation in the host cow. Based on the crude appearance of the embryoid nephrons, it is unlikely that capillary beds formed inside the Bowman's capsules to produce a glomerulus. Thus, the identity of the fluid collecting in the plastic bags is questionable.

Lanza and his associates set out not only to grow kidneys in culture but also to show the value of therapeutic cloning as applied to organ transplants. For example, a patient suffering from kidney failure currently has two options: kidney dialysis, which is costly and cumbersome, or a kidney transplant. Most kidney transplants are allografts and therefore the patient has to take immunosuppressants for the remainder of his or her life. Moreover (as shown in Table 5-1), even with immunosuppressants the long-term prognosis is not good. Therapeutic cloning offers a third alternative: Clone the patient, allow the embryo to grow long enough to harvest the stem cells, coax those cells into constructing a kidney, and then implant it in the patient. Since the kidney in this scenario is a clone of the patient, there is no threat of immune rejection. Although in theory therapeutic cloning will work, it requires the creation of a human embryo that is subsequently killed in order to provide the stem cells needed to grow the body parts. Moreover, Lanza's group used fetal kidney cells, not ES cells, and even those who may accept the destruction of a two- to three-day-old embryo to harvest stem cells balk at the idea of carrying out the same procedure on a three- or four-month-old fetus.

Society is not likely to accept the idea of using human embryos or fetuses as starter materials for growing organs, but they might accept the procedure if it started with AS, UCS, or iPS cells, particularly if they can be trained to produce real organs and not just a mass of embryoid tissue. But for this condition to be met, scientists will have to unravel the complex program that guides normal embryonic development and the formation of organs. This is a tall order, even with modern technologies, and is not likely to be achieved in the next 20 years. By that time, our understanding of basic cell biology may have advanced to a point where it will be possible to repair a defective kidney or heart without the need for organ transplants. Thus, the real value in the effort to grow artificial organs may not be the actual production of organs but the knowledge scientists gain about the control of embryogenesis and the inner workings of the cell.

Ethical Issues

Of all the new biological technologies, none have been more controversial than stem cell therapies. The only other technology to come close is animal cloning, and this is due mainly to its association with stem cells and therapeutic cloning. Ethicists and legislators have been grappling with the implications of stem cell therapy since 1998, when human embryonic stem (ES) cells were first isolated and cultured. On its face, stem cells offer the hope of curing disease and repairing a damaged nervous system. But the strategies that scientists have proposed require the destruction of human embryos and, in its most controversial form (therapeutic cloning), the creation of human embryos for the sole purpose of providing stem cells.

In the minds of many, a stem cell therapy that requires the destruction of human embryos harks back to the cruel experiments that prisoners were forced to endure at the hands of the Nazis during

World War II. This association is not as far-fetched as it may at first appear. The problems associated with stem cell research are closely related to the ethical problems affecting biomedical research in general. These issues were analyzed in the Belmont Report, a document created by the U.S. Department of Health, Education, and Welfare (HEW) (now named Health and Human Services) entitled "Ethical Principles and Guidelines for the Protection of Human Subjects of Research" in 1979 in response to abuses of basic human rights inflicted upon medical research subjects, dating back to World War II. This chapter will discuss the ethics of stem cell research in the context of that report, while leaving the legal questions to the next chapter.

JUSTICE AND BENEFICENCE

The Belmont Report is a document that provides biomedical researchers with an ethical framework to avoid the kind of abuses that occurred in Nazi Germany, and in a U.S. Public Health Service Syphilis Study at Tuskegee, Alabama. In 1932, a research institute in Tuskegee enrolled 600 low-income African-American males in a study dealing with the progression of syphilis. These patients, 400 of whom were infected with syphilis, were monitored for 40 years. The infected individuals were never told they had syphilis but instead were told that their medical problems were due to bad blood. In 1947, penicillin became widely available, and this antibiotic was known throughout the medical world as an effective treatment for syphilis. However, the participants in the Tuskegee study were never told there was an antibiotic available that could cure them. The study was terminated in 1972 by the U.S. Public Health Service only after its existence was leaked to the general public and it became a political embarrassment. By the time the study was terminated, 28 of the men had died of syphilis and 100 others were dead from related complications, 40 of the participants' wives had been infected, and 19 children had contracted the disease at birth. Public revulsion over the details of this study was instrumental in forcing the government to introduce new policies and laws regarding the use of human subjects in medical research. Four years after the termination of the study, HEW published the Belmont Report.

At the core of the report is a call for researchers to keep patients' dignity and well-being at the forefront of their concerns. The report used the word *beneficence* to capture this attitude. Beneficence generally refers to acts of kindness or charity and, in the context of the Belmont Report, is a natural extension of the Hippocratic oath that all physicians are expected to abide by: I will give no deadly medicine to anyone if asked, nor suggest any such counsel. In other words, doctors must do no harm. For those involved in biomedical research, this means never injuring one person to benefit another. An extension of the beneficence principle is the principle of justice. Researchers must never enlist subjects in an experiment if those subjects do not stand to reap any benefits. The exploitation of prisoners in Nazi concentration camps benefited the Nazis but certainly not the people experimented on. A second example cited by the report was the Tuskegee study. Aside from committing a gross deviation from the most basic of ethical standards, the designers of the Tuskegee study enlisted only black people even though they are clearly not the only racial group to suffer from this disease. The principle of justice was clearly not applied to these subjects. Their treatment was little better than what occurred in Germany's concentration camps.

To address these problems, the Belmont Report introduced, for the first time, the principle of informed consent. Physicians and biomedical researchers can enroll human subjects in a clinical research trial only if the procedure is carefully explained and the researchers receive written consent from the prospective subjects. This process was modified in 2000 to include a requirement for a patient advocate to be present when physicians are recruiting research subjects. The advocate's job is to make sure the researchers explain the procedure clearly; provide full disclosure of background research, particularly any results that may suggest the subjects will be harmed by the experiment; and, finally, ensure that there is no attempt to coerce prospective subjects into joining the trial. The inclusion of an advocate, ordered by the American Food and Drug Administration (FDA), was in response to a failed gene therapy trial in 1999 that led to the death of Jesse Gelsinger, one of the research subjects. The subsequent investigation charged the principal investigators with ignoring the safety and well-being of their subjects and with disregarding the principle of informed consent.

HIGH-TECH CANNIBALISM

Opponents of ES cell research, particularly in the form of therapeutic cloning, believe it is wrong to use human embryos for any kind of research or medical therapy and have characterized the practice as high-tech cannibalism. The issue of ES cell research is extremely complex and involves the very emotional and highly politicized issue of human abortion. Researchers want unlimited access to human embryos left over from in vitro fertilizations to use as a source of stem cells. Scientists at Advanced Cell Technology, a biotech company specializing in stem cell research, would like to push ES cell availability a step further with therapeutic cloning, whereby human embryos are created, using nuclear transfer, for the sole purpose of providing stem cells and, ultimately, culture-grown kidneys and other organs. Embryos do not survive the harvesting of the stem cells; indeed, the stem cells *are* the embryo.

Unfortunately, the Belmont Report and the Gelsinger investigation do not address these new methodologies. Nevertheless, the ethical principles now in place to guide medical research can be used to address the problems associated with ES cell research. The Belmont Report and the Gelsinger investigation established the requirement for informed consent and patient advocates. Both requirements are relevant to fetal and ES cell research. Critics argue that since neither the embryo nor the fetus is recognized as a person, in the legal sense, there is no obligation for ES cell researchers to abide by those reports. However, the absence of legal status does not preclude an ethical policy, as evidenced by the fact that abortion in North America and Europe is only permissible up to the fifth month and not beyond (except in very rare cases where the mother's life is at risk), even though the fetus does not attain the legal status of a person until after birth. With this in mind, opponents of therapeutic cloning and ES cell research maintain that at the very least a human embryo or fetus, incapable of giving informed consent, should be afforded the benefit of an advocate. Resistance to this notion centers on the question of when an embryo or a fetus becomes human.

ON BECOMING HUMAN

The question of when an embryo becomes human is not new, but one that has been debated by philosophers, scientists, and politicians for more than 2,000 years. The Greek philosopher Aristotle, always in the thick of things, believed that life arose in three stages, which he characterized as vegetative, animate, and intellectual. Aristotle was likely referring to the emergence of entire populations, but many people took it to mean that human development traveled this course as well. The embryo, immediately after conception, is in a vegetative stage, followed by an animate stage, when muscles differentiate and limbs begin to move, and finally, the individual develops an intellect. People concluded that the first two stages occurred in the womb, while the final stage appeared only after birth. Thus, in the minds of many, an embryo did not become human until after birth, and this idea has influenced public perception and dialogue ever since.

Even modern-day research scientists seem to have soaked up some of Aristotle's reasoning. Dr. John Kaplan, a noted science ethicist, has stated that cloned human embryos, destined for therapeutic research, are not really human. Other scientists, including Dr. Michael West, CEO of Advanced Cell Technology, a company that is involved in therapeutic cloning, have echoed this assessment. But sentiments expressed upon the completion of the Human Genome

Project run counter to this conclusion. At that time, many scientists declared that the human race, for the first time ever, had access to the information, contained within our genome, that defines all of our characteristics. Such statements contradict the argument that an embryo, containing a human genome, is not really human. To be sure, Kaplan and West were referring to cloned embryos, which they may think are less than human. But other scientists reject the premise that clones are less than, or even different from, their conventionally conceived genetic counterparts.

Specifically, Dr. Ian Wilmut, the British scientist who led the team that cloned Dolly the sheep, went to great lengths to assure the general public that Dolly was a sheep, like any other sheep of her breed. Wilmut, along with many other scientists, insisted that the fact that Dolly was cloned did not change her sheepness. If an embryo has a sheep's genome, it is a sheep. Following the same logic, if an embryo has a human genome, regardless of how it got it, that embryo is human. Moreover, following the logic of genomics (which states that a species identity is defined by its genes), that embryo is human from the day of conception. Many believe that it is wrong to attempt to redefine humanness for, as pointed out by the Belmont Report, such attempts have been used to justify the unethical treatment of those who are the most vulnerable, including early human embryos, which are the primary source of stem cells that researchers wish to use for medical therapies.

THE EARLY EMBRYO

ES cell research is focused primarily on the early embryo (up to 14 days old). Scientists, politicians, and the general public have been debating this technology for almost 10 years now, and some of the rancor may be traced to misconceptions regarding the development of the human embryo. Thus, any discussion of the ethical issues should be preceded by a review of the initial stages of embryonic development.

Fertilization of the female egg (oocyte) and fusion of the egg and sperm nuclei forms the zygote. The zygote undergoes a series of cell

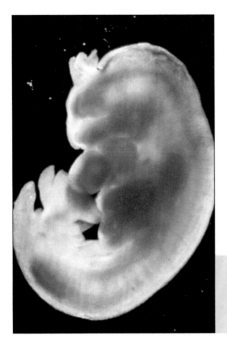

A 14-day-old human embryo.
(*Claude Edelmann/*
Photo Researchers, Inc.)

divisions beginning about 36 hours after fertilization, at which time it is referred to as an embryo. By the time the embryo is five to six days old, it consists of about 100 cells (still smaller than a period on this page) and is known as a blastocyst. The blastocyst consists of the trophoblast, which eventually forms the placenta and umbilical cord, and the inner cell mass (ICM). The ICM gives rise to the embryo and is the source of the ES cell. The blastocyst, being the source of ES cells, is the focus of much of the debate on the use of embryos in stem cell research and therapy (the rare exception involves EG cells, described in chapter 2). The blastocyst is sometimes referred to as the preimplantation embryo. About a week after fertilization implantation of the blastocyst in the womb takes place. The failure rate of this stage is high with as many as 75 percent of the blastocysts being naturally lost before implantation. At about 14 days after fertilization, following implantation, the embryo consists of about 200 cells that have begun the process of cellular differentiation. It is at this time that an anatomical feature known as the primitive streak first appears. The appearance of this feature marks the stage at which the central nervous system begins to form. For the purpose of this discussion, the term *early embryo* refers to the stages up to the appearance of the primitive streak.

The British Parliament was among the first to address the ethical and legal status of the early human embryo (the legal issues will be discussed in the next chapter). The initial intention was to regulate the practice of in vitro fertilization (IVF), pioneered by British physicians in the 1970s. Parliament established the Committee of Inquiry into Human Fertilisation and Embryology chaired by Baroness Warnock in 1982. After lengthy consultations with a broad segment of the population, the committee submitted its recommendations known as the Warnock Report in 1984. The Warnock Report recommended allowing research on early embryos but banning such experiments on embryos that are older than 14 days. With the advent of therapeutic cloning, the Parliament set up another committee in 2001 under the chairmanship of Professor Liam Donaldson to reexamine the ethical issues associated with stem cell research. The Donaldson committee called for presentations from a wide segment of society including research organizations, church groups, and individuals representing the general public, such as trade unions and the National Federation of Women's Institutes.

The central issue addressed by the Donaldson committee was the status of the early embryo. Positions stated before the committee ranged from those taken by the churches and pro-life groups that the early embryo is a human being in the fullest sense from the moment of fertilization and should be accorded the same respect as a fetus or baby, to the position that the early embryo is no more than a collection of undifferentiated cells and so deserves no more respect than any other isolated human cell or tissue. (The Warnock committee adopted a position between similar opposing views, concluding that the early embryo has a special status but not one that justifies absolute protection.)

The debate presented to the Donaldson committee focused on the principle of respect for persons. The Roman Catholic Church and pro-life groups argue that an early embryo must be treated as a person, but others, including some non-Catholic church groups, testified that the status of a person is something that develops over time and is not acquired at the point of fertilization. This view is supported by the fact that parents are more likely to grieve the loss of a newborn child than the loss, by miscarriage, of an embryo or a fetus. Moreover, there is no public mourning for the natural loss of a fetus or for surplus embryos that are disposed of at IVF clinics.

The Donaldson committee, upholding the earlier conclusions reached by the Warnock committee, recommended that the 14-day limit on embryo experimentation be maintained and that this should apply both to embryos produced by natural means and by therapeutic cloning. The committee concluded that it would be hypocritical to allow abortion but ban therapeutic cloning and early embryo experimentation. This conclusion, however, was rejected by virtually all other European countries. The European Union (EU) parliament pointed out that the issues of abortion and therapeutic cloning are separate and that one (abortion) should not be used to justify the other. Abortion is a special case in which the rights of both the mother and the fetus must be taken into account. Abortion laws in Europe and North America reflect the view that while the human fetus deserves respect, the rights of the mother are paramount. However, in the case of therapeutic cloning or IVF embryos, the question of the mothers' rights is no longer relevant and thus the fate of the embryo rests solely on the merit of its status as a developing human being.

Moreover, the EU parliament drew a clear distinction between embryos produced by therapeutic cloning and surplus IVF embryos. The IVF embryos, as acknowledged by the Donaldson committee, are produced for the purpose of creating life; that is, they are all intended for implantation and subsequent development. Therapeutic cloning creates embryos for the sole purpose of harvesting stem cells, thus killing the embryo in the process. Consequently, the EU concluded that therapeutic cloning is inherently unethical and should be banned, whereas experimentation on IVF embryos should be allowed, though tightly regulated.

The debate in the United States is more polarized and more complex. There is a greater tendency in the United States to view the banning of therapeutic cloning or ES cell research as an encroachment on the availability of abortion and the right of a woman to have one. Indeed, this perception has so far blocked the passage of laws dealing with both reproductive and therapeutic cloning in the United States, while such laws were passed in the United Kingdom and EU as far back as 2001 (discussed in the next chapter).

Polarization of the debate in the United States is due primarily to a shift in public opinion regarding the eligibility of abortion, as defined by the Supreme Court in 1973 *(Roe v. Wade)*. Public opinion polls have shown an even split between adult Americans who are for or against abortion, although many adults who believe abortion should be legal feel there should be greater restrictions and more concern for the fate of the fetus. This attitude is especially pronounced among American teenagers, the majority of whom are opposed to abortion and believe it should be illegal. Data from the Alan Guttmacher Institute has shown a steady decline in the number of abortions since the early 1990s, due primarily to a drop in teenage abortions. The drop in the abortion rate among teenagers is due to several factors. First, there has been better education among high school students, particularly regarding the biology of fetal development. This has been facilitated by the introduction of 3-d ultrasound imagery, which has produced extremely clear images of fetal development. Second, there is easier access to information about birth control and a family and social climate that encourages its use. Third, greater emphasis is being placed on the option of carrying the fetus full term for eventual adoption. Finally, the reduction in the number of abortions may also be due to reduced funding for abortion clinics and intimidation of abortion providers.

The increased concern for the fate of the embryo and fetus, combined with the ethical principles established by the Belmont Report and the Gelsinger investigation, have led to a call for embryo and fetal advocates. Parents contemplating an abortion or the possibility of donating IVF embryos for research would have access to independent advocates to discuss the issues. In the case of IVF embryos, society should not assume the clinics are acting in the best interest of the embryos any more than they can assume medical researchers are acting in the best interest of the research subjects. Indeed, in recent years, IVF clinics have been heavily criticized for creating too many embryos, many of which are never used and must then be destroyed. The clinics are profit-driven organizations that try to maximize their success rate by transplanting an excessive number of embryos (usually three or more; thus, the need for a large number of embryos). This practice often leads to multiple births, which puts the health and safety of both the mother and the infants at risk.

There are many who oppose the introduction of embryo or fetal advocates because they believe it is an attempt to undermine the legal status of abortion. However, the introduction of patient advocates, called for by the Gelsinger investigation, did not bring an end to clinical trials, nor is the introduction of fetal or embryo advocates likely to overturn abortion laws in the United States. On the contrary, fetal and embryo advocates are a natural extension of the intent of current abortion laws, both in the United States and in Europe. These laws leave the decision to terminate a pregnancy up to the parents and especially up to the mother. The fetal and embryo advocates would simply try to ensure that the parents are making an informed decision.

The United Kingdom and the EU, though divided on the issue of therapeutic cloning, have decided to allow research on human embryos, specifically those produced in IVF clinics. But it is important to note that these decisions were made without changing their abortion laws. The legislation that has been passed in the UK and EU and is now being debated before the U.S. Congress will be discussed in the next chapter.

Legal Issues

Embryonic stem (ES) cell research may provide powerful new methods for treating a variety of medical disorders. However, it also introduces many ethical problems that require legislation to control its use and spread. The legal issues deal with stem cells that are harvested from three- to five-day-old human embryos, donated by in vitro fertilization (IVF) clinics, or produced by therapeutic cloning. There are no legal issues associated with the use of adult stem (AS) cells, induced pluripotent stem (iPS) cells, or stem cells isolated from umbilical cord blood (UCS). Therapeutic cloning represents a fusion of cloning technology and stem cell research, and its regulation has proved to be an extremely difficult problem. The legal debate involving stem cells varies from country to country, particularly for therapeutic cloning. This discussion begins with the legal issues as they unfolded in the United Kingdom, for it was there that laws regulating stem cell research and therapeutic cloning were first discussed and enacted.

UNITED KINGDOM

The regulation of stem cell research in the United Kingdom is governed by the Human Fertilisation and Embryology Act of 1990. This legislation, administered by the Human Fertilisation and Embryology Authority (HFEA), was enacted to regulate the practice of IVF, which originated in Britain in 1978. The legislation was debated at length in Parliament and by the Committee of Inquiry into Human Fertilisation and Embryology. This committee was chaired by Baroness Warnock and was tabled in 1984. The Act of 1990 largely implemented the recommendations of the Warnock committee.

Under this act, research on embryos older than 14 days is prohibited. This time period was set to coincide with the appearance of the primitive streak, an anatomical feature of an embryo that indicates the beginning of neurulation and the formation of the central nervous system. All research dealing with human embryos is licensed by HFEA and may be denied if the authority feels the research objectives may be obtained with nonhuman embryos or by some other means. The license is granted only if the research is focused on the following treatments: promoting advances in the treatment for infertility, increasing knowledge about the causes of congenital disease, increasing knowledge about the causes of miscarriages, development of contraceptives, and for developing methods for detecting the presence of gene or chromosome abnormalities in embryos before implantation. There is also a provision for allowing research on embryos for "other purposes" that increase knowledge about the creation and development of embryos, as well as their potential use in developing medical therapies.

With the birth of the first mammalian clone, Dolly the sheep, in 1996, HFEA and the Human Genetics Advisory Commission undertook a public consultation on human cloning as it pertains to stem cell research. Their report, tabled in 1998, recommended that HFEA issue licenses for therapeutic cloning and that research involving the embryos so produced would be subject to the 14-day limit imposed by the act for IVF embryos. These recommendations were debated at length by the British government and passed into law as the Human Fertilisation and Embryology Regulations (HFER) on January 22, 2001. The passage of this law brought with it the concern that some of the cloned embryos might be implanted into a surrogate mother and brought to full term. To ensure that this did not happen, the government introduced the Human Reproductive Cloning bill, which proposed a ban on reproductive cloning, whereby an embryo is produced by nuclear transfer and then carried to term by a surrogate mother, as was done with Dolly. This bill was passed into law on December 4, 2001, and is still in effect.

The British legislation covering therapeutic cloning is highly regarded around the world and has served as a model for all subsequent discussions concerning cloning legislation. The legislation regulating therapeutic cloning and stem cell research (HFER) was reviewed by a special committee set up by the House of Lords in 2002 and chaired by Professor Liam Donaldson. The Donaldson committee put out a call for evidence from scientific and research organizations, churches, medical charities, patients' support groups, pro-life groups, and many organizations representing the general public, such as the trades unions and the National Federation of Women's Institutes. The committee received 52 submissions from various organizations, and they held 12 sessions of oral evidence at which 42 people representing 17 organizations presented their arguments for or against the proposed legislation. Members of the committee also visited research laboratories to gain a better understanding of the science involved.

The central issue dealt with by the Donaldson committee was the 14-day limit established by HFER. This issue was reexamined because of the great number of people and organizations that questioned the validity of the cutoff period on moral and ethical grounds. Positions ranged from the view that the early embryo is a human being in the fullest sense from the moment of fertilization and should be accorded the same protection as a human fetus or baby to the position that an early embryo is an undifferentiated collection of cells that deserves little more attention than any other isolated human cells. The Warnock committee adopted a position between these opposing views, concluding that the early embryo has a special status but not one that justifies its being accorded protection under the law. The Donaldson committee reviewed this position with respect to the principles discussed in chapter 6 (respect for persons and justice and beneficence) and on the current legal and social status of abortion. In this context, the committee focused on three main elements:

1. Legislation allowing abortion has been in place in the United Kingdom for 30 years. That legislation, known as the Abortion Act, sets an upper limit of 24 weeks for terminating a pregnancy and thus reflects a gradation in the respect accorded to a fetus as it develops from the early embryo to its birth. This does not persuade those opposed to abortion in all circumstances, but the act reflects majority public opinion and has been tested on several occasions since it was enacted. Thus, it would be difficult to justify an absolute prohibition on the destruction of early embryos while permitting abortion on fetuses long after the formation of the primitive streak.

2. IVF has been practiced in the United Kingdom for 25 years and has wide public support. As currently practiced, surplus embryos are created, most of which are eventually destroyed. To give the early embryo full protection of a person would be inconsistent with the use of IVF.

3. The 1990 act, which regulates the use of early embry-
    os for medical research, was enacted after a lengthy
    period of public and parliamentary debate and still
    retains a wide measure of public support.

Consequently, the committee concluded that the 14-day limit
was valid and should remain the limit for research on early embryos. In assigning this limit, the Warnock committee demonstrated
a respect for the early embryo that it did not previously enjoy. The
full implication of the embryo's new status was considered further
by the Donaldson committee regarding the creation of embryos for
research (therapeutic cloning). The committee received many presentations on this issue, some of which took the view that an embryo
created for research was clearly being used as a means to an end,
with no prospect of implantation, whereas embryos produced in IVF
clinics were intended for implantation, even though some would be
destroyed. On the weight of this argument the committee concluded
that embryos should not be created specifically for research purposes unless there is an exceptional need that cannot be met by the
use of surplus IVF embryos. Thus, the committee did not recom-
mend a ban on therapeutic cloning. However, it did call for increased
surveillance of the procedure to ensure that every cloned embryo is
accounted for and that the experiments to which the embryos are
subjected do not extend beyond what is allowed by law.

Critics believe the 14-day limit is nothing more than a boundary of convenience, since the legislators know that the scientists
want embryos that are less than a week old. In this view, the attempt to define a point at which a human embryo can or cannot
be destroyed is in itself unethical. HFEA's licensing decisions have
become increasingly controversial:

> ➤ In 2004 they began granting British scientists a license
>   to produce cloned human cells, making it only the sec-

ond country in the world to permit such a procedure (South Korea was the first). This is essentially a license to perform therapeutic cloning.

➤ In 2006, they approved the screening of embryos for genes that may lead to certain cancers in middle age. Such a test will encourage abortions even though the function of these genes is poorly understood.

➤ In 2007, HFEA gave women permission to donate their eggs to research projects, provided strong safeguards were put in place to ensure the women are properly informed of the risks of the procedure and are protected from coercion. Critics have argued that the amount they are offering (£250 or about $460) is an inducement that many poor women will be unable to resist and many others are likely to be coerced. HFEA insists that the women will be interviewed to ensure the donation is for altruistic reasons and not for monetary gain, but it will be impossible to know this for sure. In addition, there is a low but well-recognized risk of developing ovarian hyperstimulation syndrome, which can occur during the extraction of eggs. This syndrome can damage a woman's fertility and even cause death.

➤ In January 2008, over the objection of a large proportion of the public, HFEA granted licenses to Newcastle University and King's College London to carry out cytoplasmic hybrid research projects. These projects involve the production of human-animal hybrids, either by fusing human and animal cells or by transplanting human cells into animal embryos and allowing those animals to reach adulthood. This procedure could lead to the birth of a mouse, for example, that has a humanized brain.

Although HFEA has made several controversial decisions, the last one, regarding the production of human-animal embryos, may prove futile, as it is unlikely that these hybrid embryos will ever prove effective. Indeed, Dr. Robert Lanza and his team have recently shown that human-animal hybrid embryos are generally defective and display an abnormal gene expression profile. In particular, these hybrids fail to express critical pluripotency-associated genes, thus rendering them unusable as a source of patient-specific stem cells.

EUROPEAN UNION

The European Union (EU) agrees with the U.K.'s position on reproductive cloning and has passed laws to ban it. However, the EU strongly disagrees with the U.K. on the issue of therapeutic cloning. The Council of Europe promulgated the Convention on Human Rights and Biomedicine, which states categorically that the creation of human embryos for research purposes is prohibited. The council disagrees with the notion expressed by U.K. legislators that therapeutic cloning should be allowed simply because abortion is allowed. Abortion is a special situation in which the rights of the mother take precedence over the rights of the embryo or fetus. But when this association is broken or simply does not exist, the rights of the embryo become paramount. Thus, therapeutic cloning, or any kind of research that destroys human embryos, is illegal in Germany, Austria, Portugal, Ireland, Norway, and Poland. Even the Netherlands, a politically liberal country, passed a law in 2003 to ban the cloning of human embryos. In addition, the Council of Europe (now with 47 member states) adopted a convention on biomedicine that prohibits the creation of human embryos for research purposes. In 2002, the Netherlands and Sweden appeared willing to allow therapeutic cloning provided laws were enacted to prohibit placing such embryos in surrogate mothers to be carried to full term. But it became clear that enforcing such a law would be nearly impossible, and so a complete ban on all forms of cloning seemed to be the only practical solution.

In spring 2008, the German parliament voted to amend its stem cell law, which had restricted scientists to working on cell lines created before January 1, 2002. The amendment advances the cutoff date to May 1, 2007. It will also no longer be a criminal offense for German scientists to use even newer cell lines in countries where such research is allowed. By extending the cutoff date, legislators have given researchers access to an additional 500 cell lines. Most German scientists were hoping for a complete abandonment of the cutoff date, but the legislation has been accepted as a reasonable compromise.

UNITED STATES

A bill to prohibit all forms of cloning (Human Cloning Prohibition Act of 2001, H.R. 2505) which had the support of President George Bush, was passed by the House of Representatives in July 2001, but after seven years of debate has not as yet been written into law. The bill, introduced by Representatives David Weldon (R-Florida) and Bart Stupak (D-Michigan) had a broad base of support, but met with opposition when submitted to the Senate for debate in 2002. Dissension came from patient advocate groups, and members of the biomedical research community, who agreed to a ban on reproductive cloning but argued in favor of therapeutic cloning.

In 2003, the House of Representatives took a second vote on the bill, and this time it was approved by an overwhelming margin of 241 to 155. Senator Sam Brownback (R-Kansas) introduced the companion bill to the Senate for debate. Both bills call for a maximum penalty of $1 million in civil fines, and up to 10-year jail terms for those who attempt reproductive or therapeutic cloning. Competing legislation was also submitted by Senators Arlen Specter (then R-Pennsylvania) and Dianne Feinstein (D-California), which call for a ban only on reproductive cloning.

Legislators debating these bills before the Senate will have to deal with the same moral issues confronting the United Kingdom and EU parliaments in 2001: Namely, does a three-day-old embryo qualify for legal protection? U.K. legislation holds that an embryo less than 14 days old does qualify, whereas a 15-day-old embryo does not. The EU overwhelmingly supports a complete ban on human cloning, and it reiterated this decision in April 2003. Firm support for a total ban on human cloning from the EU, and the perceived difficulties of regulating the use of cloned embryos, will likely influence the outcome of the vote in the Senate. In addition, new research on adult stem cells and umbilical cord blood stem cells indicate that they have the same potential for curing disease as do ES cells. Thus, the argument that therapeutic cloning and ES cells are essential for the development of effective stem cell therapies is no longer convincing.

The issue of therapeutic cloning is especially difficult to resolve in the United States because the abortion issue is much more polarized in that country then it is in the United Kingdom or the EU. Antiabortion groups have seized upon the 14-day limit established by British legislators as an indication that the issue of abortion needs to be reassessed, rather than using the acceptance of abortion as an argument for allowing research on early embryos. These groups have argued that if, in the context of therapeutic cloning, it makes sense to protect a 14-day-old embryo, then it is also wrong to abort an embryo or fetus that is older than 14 days. So far, the American judiciary has refused to hear arguments of this kind, namely that a 14-day-old embryo should have the status of a person, but the various factions that are for or against this position have made it very difficult for the Congress to pass legislation even to ban reproductive cloning, a practice that is already banned in all of Europe.

In April 2004, 206 members of the House of Representatives, in response to public pressure, signed a letter urging President Bush to modify his August 2001 executive order limiting federal funds for ES cell research to preexisting cell lines. The letter called for a new policy whereby federal funds would be made available for researchers to create new ES cell lines from embryos left over from in vitro fertilization clinics. On May 14, 2004, Dr. Elias Zerhouni, the director of the National Institutes of Health, responded to the letter by reiterating President Bush's position that federal funds should not be used to "encourage further destruction of human embryos that have at least the potential for life."

Frustrated by the federal restrictions, several states held referendums to seek the public's permission to allocate state funds to establish and fund embryonic stem cell research. California, Connecticut, Illinois, Maryland, and New Jersey obtained voter approval between 2004 and 2005. At the federal level, Democrats in the House of Representatives introduced the Stem Cell Research Enhancement Act of 2005 (H.R. 810). After a lengthy debate, the bill was passed by the House and later by the Senate, but failed to get the two-thirds majority needed to block a presidential veto. President Bush, true to his word, vetoed the bill in July 2006. In his veto message, President Bush explained his objection to the bill in his opening paragraph: "Like all Americans, I believe our Nation must vigorously pursue the tremendous possibilities that science offers to cure disease and improve the lives of millions. Yet, as science brings us ever closer to unlocking the secrets of human biology, it also offers temptations to manipulate human life and violate human dignity. Our conscience and history as a Nation demand that we resist this temptation. With the right scientific techniques and the right policies, we can achieve scientific progress while living up to our ethical responsibilities."

In the following year, House Democrats reintroduced H.R. 810. They lobbied Republicans in both chambers of Congress in the hopes of winning a two-thirds majority. Although they were successful in bringing more Republicans over to their side, the bill, although passing in both chambers, failed again to get a two-thirds majority. The bill was sent to President Bush, who vetoed it on June 20, 2007, for the second time.

The second veto outraged many politicians and the scientific community as a whole. Both groups felt they commanded the moral high ground and were convinced that the public wanted the restrictions on ES cell research lifted. Pollsters have indeed reported public sympathy for the Democrat's bill, but polls can be misleading, particularly when dealing with a complex topic that is shrouded in hyperbole and misinformation. Sadly, much of the misinformation is coming from the scientists themselves who have consistently downplayed the potential of adult stem and iPS cells while leaving the impression that research in the field cannot continue without embryonic stem cells. But the facts do not support this view: To date, only one clinical trial has been launched that involves ES cells compared to more than 2,000 adult stem cell trials that are currently being funded by the National Institutes of Health (NIH). Even if scientists had an inexhaustible supply of ES cells to work with, owing to the dangers of GVHD and teratomas, they would not be able to use them in clinical trials.

Many scientists have complained bitterly that the progress of stem cell research has been hampered by President Bush's policies. This argument was in full force when it appeared that Dr. Hwang's results were reliable. American scientists were sure he had won the race because of a difference in funding. But scientists in the United Kingdom also failed to produce patient-specific ES cells, even though they have ample funding and unrestricted access to human embryos. Consequently, it may be that the pace of embryonic stem cell research in the United States has more to do with the difficulty of the problem itself and not the availability of ES cell lines or research funds. U.S. researchers have also complained that without greater access to human ES cells they will never be able to identify species-specific cellular mechanisms that control pluripotency. But researchers at Harvard University have recently completed a genome-wide analysis of mouse ES cells that has identified several mechanisms that appear to be applicable to any ES cell regardless of its source.

Nevertheless, American stem cell researchers remain committed to the goal of gaining greater access to human ES cells. On March 9, 2009, they got their wish when President Barack Obama signed an executive order reversing his predecessor's restrictions on the funding of ES cell research. However, congressional restrictions

President Barack Obama. (AP Photo/ Jessica Hill)

still remain. Although scientists can now receive government grants to do research on new stem cell lines grown in privately funded labs, they cannot use federal financing to produce new ES cell lines. Moreover, a recent Gallup poll showed that while a slight majority of the American public (52 percent) support President Obama's decision, most Americans, regardless of political affiliation or religious sentiment, want some sort of restrictions in place.

In April 2009, the Obama administration released a draft of guidelines for federal funding of human ES cell research. Under the new guidelines, federal funding will be available only for ES cells isolated from embryos created at IVF clinics for reproductive purposes. The donors would have to consent to their use for research. Funding will not be available for ES cells isolated from other sources, such as therapeutic cloning, IVF embryos created specifically for research purposes, development of an unfertilized egg, and human-animal hybrid embryos.

Although scientists and scientific societies, such as the American Society for Reproductive Medicine, have stated that the new guidelines are welcomed and long overdue, many critics believe they are unsound scientifically and fiscally irresponsible. According to Tony Perkins, president of the Family Research Council, "The NIH draft guidelines demanded by the president will do nothing to advance stem cell research that is showing near-term benefit for suffering patients. Instead of funding more embryo destructive research, the government should fund research using adult stem cells that are on the cutting edge of treating patients for diabetes, spinal cord injury, heart disease and various cancers. Unfortunately, this draft guidance only diverts limited federal resources to unethical stem cell research that has not successfully treated a single person for any disease." Thus, it would seem that the real political and social battle over this issue has only just begun.

CONCLUDING REMARKS

A casual observer may wonder why stem cells are so controversial and why the ethical and legal debate has been so protracted. After all, stem cell therapy holds the potential for curing terrible diseases such as cancer, Parkinson's disease, and Alzheimer's disease. In addition, this therapy may offer hope to those paralyzed by a spinal cord injury and someday it may be used to reverse the aging process. What could be wrong with such a wonderful procedure?

The controversy and much of the confusion center on the fact that there are three different kinds of stem cells: AS cells, ES cells, and iPS cells. AS cells are isolated from adult tissues such as bone marrow and from umbilical cord blood. ES cells are isolated from two- to five-day-old human embryos. The embryos do not survive the harvesting of the stem cells, and many people believe that it is highly immoral to kill a human embryo for its cells. The newest member of the stem cell family, iPS cells, are produced in the laboratory by reprogramming skin cells. The use of ES cells is very controversial, whereas the use of AS or iPS cells is not.

If noncontroversial stem cells are available, why not use them instead of ES cells? Why are scientists so eager to pursue ES cell research? Critics have argued that this controversy is more about politics and personal preference than it is about science.

Many scientists believe that ES cells offer the best hope for effective medical therapies because they possess a high degree of developmental plasticity (i.e., can produce a wide variety of cell types), and when injected into experimental animals they sometimes repair, or try to repair, whatever damage is present. The results of some of these experiments have been so encouraging that scientists call ES cells the gold standard of stem cell research.

But, as pointed out in previous chapters, there is a darker side to this story. The plasticity of an ES cell is in fact its Achilles' heel. Scientists may inject these cells into a mouse or a human, but they cannot control them. Studies in mice have shown that while injected ES cells initiate some repairs, they are also busy forming cancerous tumors, called teratomas, in the brain and in other parts of the body. A 17-year-old Israeli patient was injected with ES cells in 2001 to treat a neurological disorder. In 2005, after complaining about chronic headaches, he was examined by physicians at Sheba Medical Center in Tel Aviv who found tumors in his brain and spinal cord. Subsequent tests traced those tumors to the transplanted ES cells.

Fear of teratomas is the main reason why the FDA has been so reluctant to approve clinical trials involving ES cells. Of the more than 2,000 American clinical trials involving stem cells, only one is using ES cells. This one exception, pioneered by the biotech company Geron Corporation, gained approval only because the ES cells were coaxed into producing neuron precursor cells before being used, thus reducing the risk of teratoma formation. The FDA also insisted that the cultures of neuron precursors be carefully screened to ensure that they are free of undifferentiated ES cells.

If the plasticity of ES cells has to be reduced before they are useful, then surely other types of stem cells will do. AS cells are already being used to treat immune deficiencies and leukemia, and recent studies suggest that they and iPS cells will soon be used to treat spinal cord injuries, amyotrophic lateral sclerosis (ALS), kidney disease, and Parkinson's disease. Some researchers argue that AS cells are not suitable for the production of neural precursors that are needed to treat Alzheimer's or Parkinson's disease, but there is no such limitation with iPS cells.

Scientists who argue in favor of ES cell therapy have not thought it through. IPS cells can be made from skin cells taken from the patient needing treatment, and thus there is no threat of immune rejection with these cells or with AS cells. On the other hand, immune rejection is a very serious problem with ES cell therapy, since these cells are not related to the patient. The therapy can still be performed, but the patient must be kept on immunosuppressants for the rest of his or her life. This is not an acceptable solution. Even in the presence of immunosuppressants, the average five-year survival rate of conventional tissue or organ transplants is only 62 percent, dropping to 34 percent after 10 years. In the case of a conventional organ transplant, such as a heart or a lung, the patient has the option of receiving a second transplant should the original be rejected. But scientists want to use ES cell therapy to repair the brain and the spinal cord. Imagine a case where a patient suffering from Alzheimer's is successfully treated with ES cells. This means that the ES cells differentiated into the appropriate neurons and rewired the patient's brain, thus restoring his or her memories and personality. Immediately after treatment, the physicians prescribe immunosuppressants, but after five years the immune system begins to destroy the circuits produced by the ES cells. The patient is now faced with the horrible fate of developing Alzheimer's a second time.

No one knows what the outcome of the above scenario will be. Will the immune system stop after it has destroyed the neurons produced by the ES cells, or will there be collateral damage leading to the death of the patient? Will the patient be able to respond to a second round of ES cell therapy? Would it even be ethical to subject the patient to further treatment? Keep in mind that in this case the patient's mental faculties would be so badly damaged that someone else would have to make these decisions. Viewed in this way, ES cell therapy becomes a classic example of short-term gain for long-term pain.

So let us consider the alternative scenario: A patient suffering from Alzheimer's is treated with AS or iPS cells. Critics would argue that these cells would not be as effective as ES cells. For the sake of argument, let us assume that they are right. The treatment does not fully restore the patient to health. But even a 30 percent improvement in their cognitive skills would make a huge difference in the everyday life of such a patient. They may not be able to remember everything, but they could function again, they could regain their independence, and they could once again recognize the faces of their loved ones. Above all, they would know that the gift of a new life would not be torn away from them five years down the road. Sometimes an imperfect therapy is the best one of all.

10

Resource Center

Stem cell research would not be possible without the knowledge and procedures that have been provided by modern cell biology and biotechnology, both of which are discussed in this chapter. Additional resource material is provided, including an introduction to gene therapy, a brief overview of the Belmont Report, a summary of the investigation into the death of Jesse Gelsinger, and the design of clinical trials.

CELL BIOLOGY

A cell is a microscopic life-form made from a variety of nature's building blocks. The smallest of these building blocks are subatomic particles known as quarks and leptons that form protons, neutrons, and electrons, which in turn form atoms. Scientists have identified more than 200 atoms, each of which represents a fundamental element of nature; carbon, oxygen, and nitrogen are common examples.

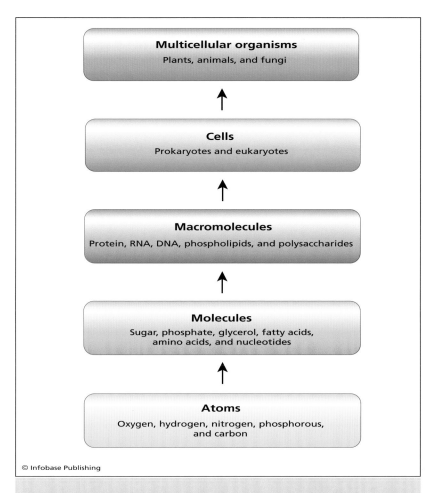

Nature's building blocks. Particles known as quarks and leptons, created in the heat of the big bang, formed the first atoms, which combined to form molecules in the oceans of the young Earth. Heat and electrical storms promoted the formation of macromolecules, providing the building blocks for cells, which in turn went on to form multicellular organisms.

Atoms, in their turn, can associate with each other to form another kind of building block known as a molecule. Sugar, for example, is a molecule constructed from carbon, oxygen, and hydrogen, while ordinary table salt is a molecule consisting of sodium and chloride. Molecules can link up with one another to form yet another kind of building block known as a macromolecule. Macromolecules, present in the atmosphere of the young Earth, gave rise to cells, which in turn went on to form multicellular organisms; in forming those organisms, cells became a new kind of building block.

The Origin of Life

Molecules essential for life are thought to have formed spontaneously in the oceans of the primordial Earth about 4 billion years ago. Under the influence of a hot stormy environment, the molecules combined to produce macromolecules, which in turn formed microscopic bubbles that were bounded by a sturdy macromolecular membrane analogous to the skin on a grape. It took about half a billion years for the prebiotic bubbles to evolve into the first cells, known as prokaryotes, and another 1 billion years for those cells to evolve into the eukaryotes. Prokaryotes, also known as bacteria, are small cells (about five micrometers in diameter) that have a relatively simple structure and a genome consisting of about 4,000 genes. Eukaryotes are much larger (about 30 micrometers in diameter), with a complex internal structure and a very large genome, often exceeding 20,000 genes. These genes are kept in a special organelle called the nucleus (eukaryote means "true nucleus"). Prokaryotes are all single-cell organisms, although some can form short chains or temporary fruiting bodies. Eukaryotes, on the other hand, gave rise to all of the multicellular plants and animals that now inhabit the Earth.

A Typical Eukaryote

Eukaryotes assume a variety of shapes that are variations on the simple spheres from which they originated. Viewed from the side, they often have a galactic profile, with a central bulge (the nucleus), tapering to a thin disclike shape at the perimeter. The internal structure is complex, being dominated by a large number of organelles.

The functional organization of a eukaryote is analogous to a carpentry shop, which is usually divided into two main areas—the

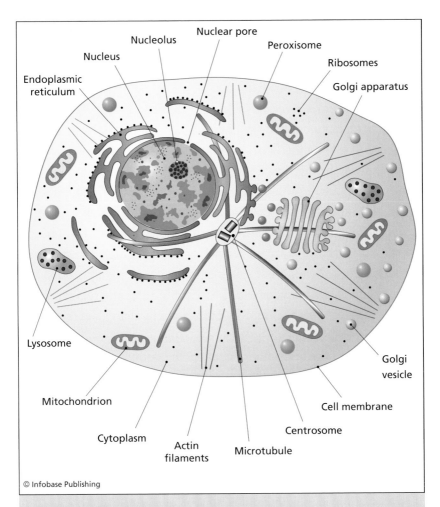

The eukaryote cell. The structural components shown here are present in organisms as diverse as protozoans, plants, and animals. The nucleus contains the DNA genome and an assembly plant for ribosomal subunits (the nucleolus). The endoplasmic reticulum (ER) and the Golgi work together to modify proteins, most of which are destined for the cell membrane. These proteins travel from the ER to the Golgi and from the Golgi to their final destination in transport vesicles (red and yellow spheres). Mitochondria provide the cell with energy in the form of adenosine triphospate (ATP). Ribosomes, some of which are attached to the ER, synthesize proteins. Lysosomes and peroxisomes recycle cellular material. The microtubules and centrosome form the spindle apparatus for moving chromosomes to the daughter cells during cell division. Actin and other protein filaments form a weblike cytoskeleton.

shop floor where the machinery, building materials, and finishing rooms are kept and the shop office where the work is coordinated and the blueprints are stored for everything the shop makes. Carpentry shops keep a blueprint on file for every item that is made. When the shop receives an order, perhaps for a chair, someone in the office makes a copy of the chair's blueprint and delivers it to the carpenters on the shop floor. In this way, the master copy is kept out of harm's way, safely stored in the filing cabinet. The carpenters, using the blueprint copy and the materials and tools at hand, build the chair, and then they send it into a special room where it is painted. After the chair is painted, it is taken to another room where it is polished and then packaged for delivery. The energy for all of this activity comes through the electrical wires that are connected to a power generator somewhere in the local vicinity. The shop communicates with other shops and its customers by using the telephone, e-mail, or postal service.

In the cell, the shop floor is called the cytoplasm and the shop office is the nucleus. Eukaryotes make a large number of proteins, and they keep a blueprint for each one. In this case, however, the blueprints are not pictures on pieces of paper, but molecules of deoxyribonucleic acid (DNA) that are kept in the nucleus. A cellular blueprint is called a gene, and a typical cell has thousands of them. A human cell, for example, has 30,000 genes, all of which are kept on 46 separate DNA molecules known as chromosomes (23 from each parent). When the cell decides to make a protein, it begins by making a ribonucleic acid (RNA) copy of the protein's gene. This blueprint copy, known as messenger RNA (mRNA), is made in the nucleus and delivered to the cell's carpenters in the cytoplasm. These carpenters are enzymes that control and regulate all of the cell's chemical reactions. Some of the enzymes are part of a complex protein-synthesizing machine known as a ribosome. Cytoplasmic enzymes and the ribosomes synthesize proteins using mRNA as the template, after which many of the proteins are sent to a compartment known as the endoplasmic reticulum (ER), where they are glycosylated or "painted" with sugar molecules. From there, they are shipped to another compartment called the Golgi apparatus, where the glycosylation is refined before the finished products, now looking like molecular trees, are loaded into transport bubbles and shipped to their final destination.

The shape of the cell is maintained by an internal cytoskeleton composed of actin and intermediate filaments. Mitochondria, once free-living prokaryotes, provide the cell with energy in the form of adenosine triphosphate (ATP). The production of ATP is carried out by an assembly of metal-containing proteins, called the electron transport chain, located in the mitochondrion inner membrane. Lysosomes and peroxisomes process and recycle cellular material and molecules. The cell communicates with other cells and the outside world through a forest of glycoproteins, known as the glycocalyx, that covers the cell surface. Producing and maintaining the glycocalyx is the principal function of the ER and Golgi apparatus and a major priority for all eukaryotes.

Cells are biochemical entities that synthesize many thousands of molecules. Studying these chemicals and the biochemistry of the cell would be extremely difficult were it not for the fact that most of the chemical variation is based on six types of molecules that are assembled into just five types of macromolecules. The six basic molecules are amino acids, phosphate, glycerol, sugars, fatty acids, and nucleotides. The five macromolecules are proteins, DNA, RNA, phospholipids, and sugar polymers called polysaccharides.

Molecules of the Cell

Amino acids have a simple core structure consisting of an amino group, a carboxyl group, and a variable R group attached to a carbon atom. There are 20 different kinds of amino acid, each with a unique R group. The simplest and most ancient amino acid is glycine, with an R group that consists only of hydrogen. The chemistry of the various amino acids varies considerably: Some carry a positive electric charge, while others are negatively charged or electrically neutral; some are water soluble (hydrophilic), while others are hydrophobic.

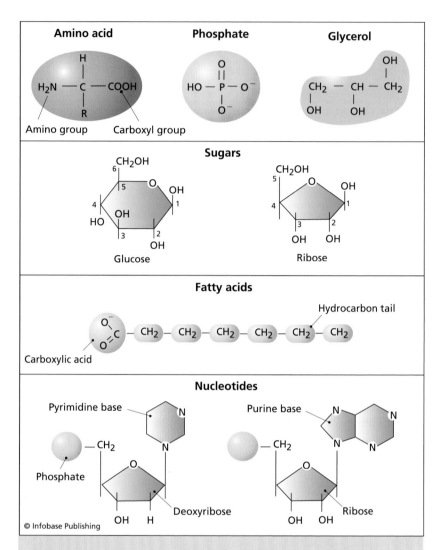

Molecules of the cell. Amino acids are the building blocks for proteins. Phosphate is an important component of many other molecules and is added to proteins to modify their behavior. Glycerol is an alcohol that is an important ingredient in cell membranes and fat. Sugars, such as glucose, are a primary energy source for most cells and also have many structural functions (the carbon atoms are numbered). Fatty acids are involved in the production of cell membranes and storage of fat. Nucleotides are the building blocks for DNA and RNA: P: phosphate; C: carbon; H: hydrogen; O: oxygen; N: nitrogen; and R: variable molecular group.

Phosphates are extremely important molecules that are used in the construction or modification of many other molecules. They are also used to store chemical-bond energy in the form of ATP. The production of phosphate-to-phosphate chemical bonds for use as an energy source is an ancient cellular process, dating back at least 2 billion years.

Glycerol is a simple three-carbon alcohol that is an important component of cell membranes and fat reservoirs. This molecule may have stabilized the membranes of prebiotic bubbles. Interestingly, it is often used today as an ingredient for making long-lasting soap bubbles.

Sugars are versatile molecules, belonging to a general class of compounds known as carbohydrates that serve a structural role as well as providing energy for the cell. Glucose, a six-carbon sugar, is the primary energy source for most cells and the principal sugar used to glycosylate the proteins and lipids that form the outer coat of all cells. Plants have exploited the structural potential of sugars in their production of cellulose; wood, bark, grasses, and reeds are all polymers of glucose and other monosaccharides. Ribose, a five-carbon sugar, is a component of nucleic acids as well as the cell's main energy depot, ATP. The numbering convention for sugar carbon atoms is shown in the figure above. Ribose carbons are numbered as 1' (1 prime), 2', and so on. Consequently, references to nucleic acids, which include ribose, often refer to the 3' or 5' carbon.

Fatty acids consist of a carboxyl group (the hydrated form is called carboxylic acid) linked to a hydrophobic hydrocarbon tail. These molecules are used in the construction of cell membranes and fat. The hydrophobic nature of fatty acids is critically important to the normal function of the cell membrane since it prevents the passive entry of water and water-soluble molecules.

Nucleotides are building blocks for DNA and RNA. These molecules consist of three components: a phosphate, a ribose sugar, and a nitrogenous (nitrogen-containing) ring compound that behaves as a base in solution (a base is a substance that can accept a proton in solution). Nucleotide bases appear in two forms: a single-ring nitrogenous base called a pyrimidine and a double-ringed base called a purine. There are two kinds of purines (adenine and guanine) and three pyrimidines (uracil, cytosine, and thymine). Uracil is specific to RNA, substituting for thymine. In addition, RNA nucleotides contain ribose, whereas DNA nucleotides contain deoxyribose (hence the origin of their names). Ribose has a hydroxyl (OH) group attached to both the 2' and 3' carbons, whereas deoxyribose is missing the 2' hydroxyl group.

Macromolecules of the Cell

The six basic molecules are used by all cells to construct five essential macromolecules: proteins, RNA, DNA, phospholipids, and polysaccharides. Macromolecules have primary, secondary, and tertiary structural levels. The primary structural level refers to the chain that is formed by linking the building blocks together. The secondary structure involves the bending of the linear chain to form a three-dimensional object. Tertiary structural elements involve the formation of chemical bonds between some of the building blocks in the chain to stabilize the secondary structure. A quaternary structure can also occur when two identical molecules interact to form a dimer or double molecule.

Proteins are long chains or polymers of amino acids. The primary structure is held together by peptide bonds that link the carboxyl end of one amino acid to the amino end of a second amino acid. Thus, once constructed, every protein has an amino end and a carboxyl end. An average protein consists of about 400 amino acids. There are 21 naturally occurring amino acids; with this number the cell can produce an almost infinite variety of proteins. Evolution and natural selection, however, have weeded out most of these, so that eukaryote cells function well with 10,000 to 30,000 different proteins. In addition, this select group of proteins has been conserved over the past 2 billion years (i.e., most of the proteins found in yeast can also be found, in modified form, in humans and other higher organisms). The secondary structure of a protein depends on the

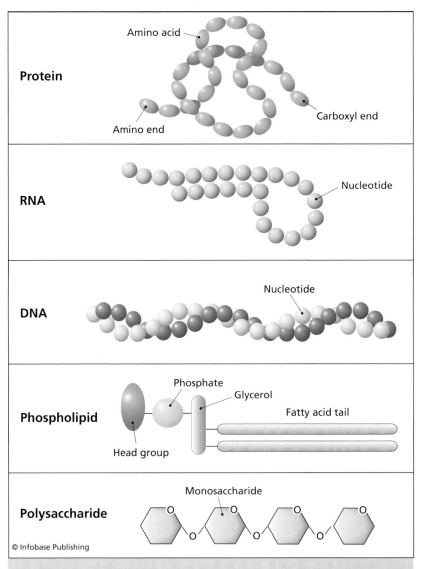

Protein

Amino acid

Amino end

Carboxyl end

RNA

Nucleotide

DNA

Nucleotide

Phospholipid

Phosphate

Glycerol

Fatty acid tail

Head group

Polysaccharide

Monosaccharide

© Infobase Publishing

Macromolecules of the cell. Protein is made from amino acids linked together to form a long chain that can fold up into a three-dimensional structure. RNA and DNA are long chains of nucleotides. RNA is generally single stranded but can form localized double-stranded regions. DNA is a double-stranded helix, with one strand coiling around the other. A phospholipid is composed of a hydrophilic headgroup, a phosphate, a glycerol molecule, and two hydrophobic fatty acid tails. Polysaccharides are sugar polymers.

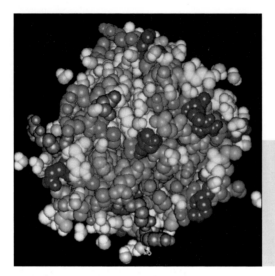

Computer model of a Ras protein. This is involved in cancer cell formation. (James King-Holmes/Photo Researchers, Inc.)

amino acid sequence and can be quite complicated, often producing three-dimensional structures possessing multiple functions.

RNA is a polymer of the ribonucleotides adenine, uracil, cytosine, and guanine. RNA is generally single-stranded, but it can form localized double-stranded regions by a process known as complementary base pairing, whereby adenine forms a bond with uracil and cytosine pairs with guanine. RNA is involved in the synthesis of proteins and is a structural and enzymatic component of ribosomes.

DNA is a double-stranded nucleic acid. This macromolecule encodes cellular genes and is constructed from adenine, thymine, cytosine, and guanine deoxyribonucleotides. The two DNA strands coil around each other like strands in a piece of rope, creating a double helix. The two strands are complementary throughout the length of the molecule—adenine pairs with thymine and cytosine pairs with guanine. Thus, if the sequence of one strand is known to be ATCGTC, the sequence of the other strand must be TAGCAG.

Phospholipids are the main component in cell membranes. These macromolecules are composed of a polar head group (usually an alcohol), a phosphate, glycerol, and two hydrophobic fatty acid tails. Fat that is stored in the body as an energy reserve has a structure similar to a phospholipid, being composed of three fatty acid chains attached to a molecule of glycerol. The third fatty acid takes the place of the phosphate and head group of a phospholipid.

Polysaccharides are sugar polymers consisting of two or more monosaccharides. Disaccharides (two monosaccharides) and oligosaccharides (about three to 12 monosaccharides) are attached to proteins and lipids destined for the cell surface or the extracellular matrix. Polysaccharides such as glycogen and starch may contain several hundred monosaccharides and are stored in cells as an energy reserve.

Basic Cellular Functions

There are six basic cellular functions: DNA replication, DNA maintenance, gene expression, power generation, cell division, and cell

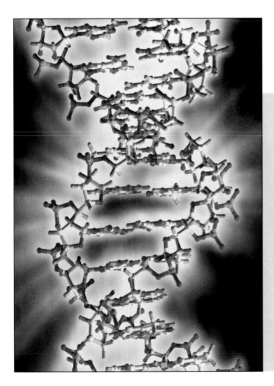

Computer artwork of a strand of DNA on a colored background. Atoms are depicted as sticks and balls and are color coded: carbon (white), oxygen (red), hydrogen (pale blue), nitrogen (dark blue), and phosphorous (orange). DNA spirals clockwise when seen from above. (Alfred Pasieka/Photo Researchers, Inc.)

communication. DNA replication usually occurs in conjunction with cell division, but there are exceptions known as polyploidization (see the glossary). Gene expression refers to the process whereby the information stored in a gene is used to synthesize RNA or protein. The production of power is accomplished by extracting energy from food molecules and then storing that energy in a form that is readily available to the cell. Cells communicate with their environment and with other cells. The communication hardware consists of a variety of special macromolecules that are embedded in the cell membrane.

DNA Replication

Replication is made possible by the complementarity of the two DNA strands. Since adenine (A) always pairs with thymine (T), and guanine (G) always pairs with cytosine (C), replication enzymes are able to duplicate the molecule by treating each of the original strands as templates for the new strands. For example, if a portion of the template strand reads ATCGTTGC, the new strand will be TAGCAACG.

DNA replication requires the coordinated effort of a team of enzymes, led by DNA helicase and primase. The helicase separates the two DNA strands at the astonishing rate of 1,000 nucleotides every second. This enzyme gets its name from the fact that it unwinds the DNA helix as it separates the two strands. The enzyme that is directly responsible for reading the template strand and for synthesizing the new daughter strand is called DNA polymerase. This enzyme also has an editorial function; it checks the preceding nucleotide to make sure it is correct before it adds a nucleotide to the growing chain. The editor function of this enzyme introduces an interesting problem. How can the polymerase add the very first nucleotide, when it has to check a preceding nucleotide before adding a new one? A special enzyme, called primase, which is attached to the helicase, solves this problem. Primase synthesizes short pieces of RNA that form a DNA-RNA double-stranded region. The

RNA becomes a temporary part of the daughter strand, thus priming the DNA polymerase by providing the crucial first nucleotide in the new strand. Once the chromosome is duplicated, DNA repairs enzymes, discussed below, removes the RNA primers, and replaces them with DNA nucleotides.

DNA Maintenance

Every day, in a typical human cell, thousands of nucleotides are being damaged by spontaneous chemical events, environmental pollutants, and radiation. In many cases, it takes only a single defective nucleotide within the coding region of a gene to produce an inactive, mutant protein. The most common forms of DNA damage are depurination and deamination. Depurination is the loss of a purine base (guanine or adenine) resulting in a gap in the DNA sequence, referred to as a missing tooth. Deamination converts cytosine to uracil, a base that is normally found only in RNA.

About 5,000 purines are lost from each human cell every day, and over the same time period 100 cytosines are deaminated per cell. Depurination and deamination produce a great deal of damage and, in either case, the daughter strand ends up with a missing nucleotide and possibly a mutated gene, as the DNA replication machinery simply bypasses the uracil or the missing tooth. If left unrepaired, the mutated genes will be passed on to all daughter cells, with catastrophic consequences for the organism as a whole.

DNA damage caused by depurination is repaired by special nuclear proteins that detect the missing tooth, excise about 10 nucleotides on either side of the damage, and then, using the complementary strand as a guide, reconstruct the strand correctly. Deamination is dealt with by a special group of DNA repair enzymes known as base-flippers. These enzymes inspect the DNA one nucleotide at a time. After binding to a nucleotide, a base-flipper breaks the hydrogen bonds holding the nucleotide to its complementary partner. It then performs the maneuver for which it gets its name. Holding onto the nucleotide, it rotates the base a full 180 degrees, inspects it carefully, and, if it detects any damage, cuts the base out and discards it. In this case, the base- flipper leaves the final repair to the missing-tooth crew that detects and repairs the gap as described previously. If the nucleotide is normal, the base-flipper rotates it back into place and reseals the hydrogen bonds. Scientists have estimated that these maintenance crews inspect and repair the entire genome of a typical human cell in less than 24 hours.

Gene Expression

Genes encode proteins and several kinds of RNA. Extracting the coded information from DNA requires two sequential processes known as transcription and translation. A gene is said to be expressed when either or both of these processes have been completed. Transcription, catalyzed by the enzyme RNA polymerase, copies one strand of the DNA into a complementary strand of mRNA, which is sent to the cytoplasm, where it joins with a ribosome. Translation is a process that is orchestrated by the ribosomes. These particles synthesize proteins using mRNA and the genetic code as guides. The ribosome can synthesize any protein specified by the mRNA, and the mRNA can be translated many times before it is recycled. Some RNAs, such as ribosomal RNA and transfer RNA are never translated. Ribosomal RNA (rRNA) is a structural and enzymatic component of ribosomes. Transfer RNA (tRNA), though separate from the ribosome, is part of the translation machinery.

The genetic code provides a way for the translation machinery to interpret the sequence information stored in the DNA molecule and represented by mRNA. DNA is a linear sequence of four different kinds of nucleotides, so the simplest code could be one in which each nucleotide specifies a different amino acid; that is, adenine coding for the amino acid glycine, cytosine for lysine, and so on. The earliest cells may have used this coding system, but it is limited to the construction of proteins consisting of only four different kinds of amino acids. Eventually, a more elaborate code evolved in which a combination of three out of the four possible DNA nucleotides, called codons, specifies a single amino acid. With this scheme, it is possible to have a unique code for each of the 20 naturally occurring amino acids. For example, the codon AGC specifies the amino acid serine, whereas TGC specifies the amino acid cysteine. Thus, a gene may be viewed as a long continuous sequence of codons. However, not all codons specify an amino acid. The sequence TGA signals the end of the gene, and a special codon, ATG, signals the start site, in addition to specifying the amino acid methionine. Consequently, all proteins begin with this amino acid, although it is sometimes removed once construction of the protein is complete. As mentioned above, an average protein may consist of 300 to 400 amino acids; since the codon consists of three nucleotides for each amino acid, a typical gene may be 900 to 1,200 nucleotides long.

Power Generation

Dietary fats, sugars, and proteins, not targeted for growth, storage, or repairs, are converted to ATP by the mitochondria. This process requires a number of metal-binding proteins, called the respiratory chain (also known as the electron transport chain), and a special ion channel-enzyme called ATP synthase. The respiratory chain consists of three major components: NADH dehydrogenase, cytochrome b, and cytochrome oxidase. All of these components are protein complexes with an iron (NADH dehydrogenase, cytochrome b) or a copper core (cytochrome oxidase), and together with the ATP synthase, are located in the inner membrane of the mitochondria.

The respiratory chain is analogous to an electric cable that transports electricity from a hydroelectric dam to homes. The human body, like that of all animals, generates electricity by processing food molecules through a metabolic pathway called the Krebs cycle, also located within the mitochondria. The electrons (electricity) so generated are transferred to hydrogen ions, which quickly bind to a special nucleotide called nicotinamide adenine dinucleotide (NAD). Binding of the hydrogen ion to NAD is noted by abbreviating the resulting molecule as NADH. The electrons begin their journey down the respiratory chain when NADH binds to NADH dehydrogenase, the first component in the chain. This enzyme does just what its name implies: It removes the hydrogen from NADH, releasing the stored electrons, which are conducted through the chain by the iron and copper as though they were traveling along an electric wire. As the electrons travel from one end of the chain to the other, they energize the synthesis of ATP, which is released from the mitochondria for use by the cell. All electrical circuits must have a ground, that is, the electrons need someplace to go once they have completed the circuit. In the case of the respiratory chain, the ground is oxygen. After passing through cytochrome oxidase, the last component in the chain, the electrons are picked up by oxygen, which combines with hydrogen ions to form water.

The Cell Cycle

Free-living single cells divide as a way of reproducing their kind. Among plants and animals, cells divide as the organism grows from a seed, or an embryo, into a mature individual. This form of cell division, in which the parent cell divides into two identical daughter cells, is called mitosis. A second form of cell division, known as meiosis, is intended for sexual reproduction and occurs exclusively in gonads.

Cell division is part of a grander process known as the cell cycle, which consists of two phases: interphase and M phase (meiosis or mitosis). Interphase is divided into three subphases called Gap 1 (G_1), S phase (a period of DNA synthesis), and Gap 2 (G_2). The conclusion of interphase, and with it the termination of G_2, occurs with division of the cell and a return to G_1. Cells may leave the cycle by entering a special phase called G_0. Some cells, such as post-mitotic neurons in an animal's brain, remain in G_0 for the life of the organism. For most cells, the completion of the cycle, known as the generation time, can take 30 to 60 minutes.

Cells grow continuously during interphase while preparing for the next round of division. Two notable events are the duplication of the spindle (the centrosome and associated microtubules), a struc-

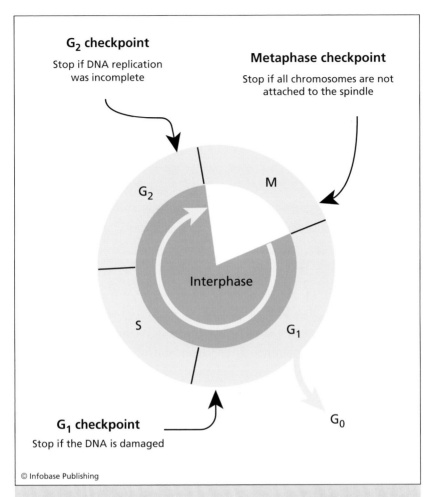

G₂ checkpoint

Stop if DNA replication
was incomplete

Metaphase checkpoint

Stop if all chromosomes are not
attached to the spindle

G₂

M

Interphase

S

G₁

G₁ checkpoint

Stop if the DNA is damaged

G₀

© Infobase Publishing

The cell cycle. Many cells spend their time cycling between inter-phase and M phase (cell division by mitosis or meiosis). Interphase is divided into three subphases: Gap 1(G₁), S phase (DNA synthesis), and Gap 2 (G₂). Cells may exit the cycle by entering G₀. The cell cycle is equipped with three checkpoints to ensure that the daughter cells are identical and that there is no genetic damage. The yellow arrow indicates the direction of the cycle.

ture that is crucial for the movement of the chromosomes during cell division, and the appearance of an enzyme called maturation promoting factor (MPF) at the end of G_2. MPF phosphorylates histones, proteins that bind to the DNA, and when phosphorylated compact (or condense) the chromosomes in preparation for cell division. MPF is also responsible for the breakdown of the nuclear membrane. When cell division is complete, MPF disappears, allowing the chromosomes to decondense and the nuclear envelope to reform. Completion of a normal cell cycle always involves the division of a cell into two daughter cells, either meiotically or mitotically.

Cell division is such a complex process that many things can, and do, go wrong. Cell cycle monitors, consisting of a team of enzymes, check to make sure that everything is going well each time a cell divides, and, if it is not, those monitors stop the cell from dividing until the problem is corrected. If the damage cannot be repaired, a cell remains stuck in midstream for the remainder of its life. If this happens to a cell in an animal's body, it is forced to commit suicide, in a process called apoptosis, by other cells in the immediate neighborhood or by the immune system.

The cell cycle includes three checkpoints: The first is a DNA damage checkpoint that occurs in G_1. The monitors check for damage that may have occurred as a result of the last cell cycle or were caused by something in the environment, such as UV radiation or toxic chemicals. If damage is detected, DNA synthesis is blocked until it can be repaired. The second checkpoint occurs in G_2, where the monitors make sure errors were not introduced when the chromosomes were duplicated during S-phase. The G_1 and G_2 checkpoints are sometimes referred to collectively as DNA damage checkpoints. The third and final checkpoint occurs in M-phase, to ensure that all of the chromosomes are properly attached to the spindle. This checkpoint is intended to prevent gross abnormalities in the daughter cells with regard to chromosome number. If a chromosome fails to attach to the spindle, one daughter cell will end up with too many chromosomes, while the other will have too few.

Mitosis

Mitosis is divided into four stages known as prophase, metaphase, anaphase, and telophase. The behavior and movement of the chromosomes characterize each stage. At prophase, DNA replication has already occurred, and the nuclear membrane begins to break down. Condensation of the duplicated chromosomes initiates the phase (i.e., the very long, thin chromosomes are folded up to produce short, thick chromosomes that are easy to move and maneuver). Under the microscope, the chromosomes become visible as X-shaped structures, which are the two duplicated chromosomes, often called sister chromatids. A special region of each chromosome, called a centromere, holds the chromatids together. Proteins bind to the centromere to form a structure called the kinetochore. The centrosome is duplicated, and the two migrate to opposite ends of the cell.

During metaphase, the chromosomes are sorted out and aligned between the two centrosomes. By this time, the nuclear membrane has completely broken down. The two centrosomes and the microtubules fanning out between them form the mitotic spindle. The area in between the spindles, where the chromosomes are aligned, is known as the metaphase plate. Some of the microtubules make contact with the kinetochores, while others overlap, with motor proteins situated in between.

Anaphase begins when the duplicated chromosomes move to opposite poles of the cell. The first step is the release of an enzyme that breaks the bonds holding the kinetochores together, thus allowing the sister chromatids to separate from one aother while remaining bound to their respective microtubules. Motor proteins, using energy supplied by ATP, move along the microtubule, dragging the chromosomes to opposite ends of the cell.

During telophase, the daughter chromosomes arrive at the spindle poles and decondense to form the relaxed chromosomes characteristic of interphase nuclei. The nuclear envelope begins forming around the chromosomes, marking the end of mitosis. By the end of telophase, individual chromosomes are no longer distinguishable and are referred to as chromatin. While the nuclear membrane reforms, a contractile ring, made of the proteins myosin and actin, begins pinching the parental cell in two. This stage, separate from mitosis, is called cytokinesis and leads to the formation of two daughter cells, each with one nucleus.

Meiosis

Many eukaryotes reproduce sexually through the fusion of gametes (eggs and sperm). If gametes were produced mitotically, a catastrophic growth in the number of chromosomes would occur each time a sperm fertilized an egg. Meiosis is a special form of cell division that prevents this from happening by producing haploid gametes, each possessing half as many chromosomes as the diploid cell. When haploid gametes fuse, they produce an embryo with the correct number of chromosomes.

Unlike mitosis, which produces two identical daughter cells, meiosis produces four genetically unique daughter cells that have half the number of chromosomes found in the parent cell. This is possible because meiosis consists of two rounds of cell division, called meiosis I and meiosis II, with only one round of DNA synthesis. Microbiologists discovered meiosis almost 100 years ago by comparing the number of chromosomes in somatic cells and germ cells. The roundworm, for example, was found to have four chromosomes in its somatic cells but only two in its gametes. Many other studies also compared the amount of DNA in nuclei from somatic cells and gonads, always with the same result: The amount of DNA in somatic cells is at least double the amount in fully mature gametes.

Meiotic divisions are divided into the four mitotic stages discussed above. Indeed, meiosis II is virtually identical to a mitotic division. Meiosis I resembles mitosis, but close examination shows two important differences: Gene swapping occurs between homologous chromosomes in prophase, producing recombinant chromosomes, and maternal and paternal chromosomes are distributed to different daughter cells. At the end of meiosis I, one of the daughter cells contains a mixture of normal and recombinant maternal chromosomes and the other contains normal and recombinant paternal chromosomes. During meiosis II, the duplicated chromosomes are distributed to different daughter cells, yielding four genetically unique cells: paternal, paternal recombinant, maternal, and maternal recombinant. Mixing genetic material in this way is unique to meiosis, and it is one of the reasons sexual reproduction has been such a powerful evolutionary force.

Cell Communication

A forest of glycoproteins and glycolipids covers the surface of every cell like trees on the surface of the Earth. The cell's forest is called the glycocalyx, and many of its trees function like sensory antennae. Cells use these antennae to communicate with their environment and with other cells. In multicellular organisms, the glycocalyx also plays an important role in holding cells together. In this case, the antennae of adjacent cells are connected to one another through the formation of chemical bonds.

The sensory antennae, also known as receptors, are linked to a variety of secondary molecules that serve to relay messages to the interior of the cell. These molecules, some of which are called second messengers, may activate machinery in the cytoplasm or may enter the nucleus to activate gene expression. The signals that a cell receives are of many different kinds but generally fall into one of five categories: 1) proliferation, which stimulates the cell to grow and divide; 2) activation, which is a request for the cell to synthesize and release specific molecules; 3) deactivation, which serves as a brake for a previous activation signal; 4) navigation, which helps direct the cell to a specific location (this is very important for free-living cells hunting for food and for immune system cells that are hunting for invading microorganisms); 5) termination, which is a signal that orders the cell to commit suicide. This death signal occurs during embryonic development (e.g., the loss of webbing between the fingers and toes) and during an infection. In some cases, the only way the immune system can deal with an invading pathogenic microbe is to order some of the infected cells to commit suicide. This process is known as apoptosis.

BIOTECHNOLOGY

Biotechnology (also known as recombinant DNA technology) consists of several procedures that are used to study the structure and function of genes and their products. Central to this technology is the ability to clone specific pieces of DNA and to construct libraries of these DNA fragments that represent the genetic repertoire of an entire organism or a specific cell type. With these libraries at hand, scientists have been able to study the cell and whole organisms in unprecedented detail. The information so gained has revolutionized biology as well as many other disciplines, including medical science, pharmacology, psychiatry, and anthropology, to name but a few.

DNA Cloning

In 1973, scientists discovered that restriction enzymes (enzymes that can cut DNA at specific sites), DNA ligase (an enzyme that can join two pieces of DNA together), and bacterial plasmids could be used to clone DNA molecules. Plasmids are small (about 3,000 base pairs) circular mini-chromosomes that occur naturally in bacteria and are often exchanged between cells by passive diffusion. A bacterium is said to be transfected when it acquires a new plasmid. For bacteria, the main advantage to swapping plasmids is that they often carry antibiotic resistance genes, so that a cell sensitive to ampicillin can become resistant simply by acquiring the right plasmid. For scientists, plasmid-swapping provided an ideal method for amplifying or cloning a specific piece of DNA.

The first cloning experiment used a plasmid from the bacterium *Escherichia coli* that was cut with the restriction enzyme *Eco*RI. The plasmid had a single *Eco*RI site so the restriction enzyme simply opened the circular molecule. Foreign DNA, cut with the same restriction enzyme, was incubated with the plasmid. Because the plasmid and foreign DNA were both cut with *Eco*RI, the DNA could insert itself into the plasmid to form a hybrid, or recombinant plasmid, after which DNA ligase sealed the two together. The reaction mixture was added to a small volume of *E. coli* so that some of the cells could take up the recombinant plasmid before being transferred to a nutrient broth containing streptomycin. Only those cells carrying the recombinant plasmid, which contained an anti-streptomycin gene, could grow in the presence of this antibiotic. Each time the cells divided, the plasmid DNA was duplicated along with the main chromosome. After the cells had grown overnight, the foreign DNA had been amplified billions of times and was easily isolated for sequencing or expression studies. In this procedure, the plasmid is known as a cloning vector because it serves to transfer the foreign DNA into a cell.

DNA Libraries

The basic cloning procedure described above not only provides a way to amplify a specific piece of DNA but can also be used to construct DNA libraries. In this case, however, the cloning vector is a bacterio-phage called lambda. The lambda genome is double-stranded DNA of about 40,000 base pairs (bp), much of which can be replaced by foreign DNA without sacrificing the ability of the virus to infect bacteria. This is the great advantage of lambda over a plasmid. Lambda can accommodate very long pieces of DNA, often long enough to contain an entire gene, whereas a plasmid cannot accommodate for-eign DNA that is larger than 2,000 base pairs. Moreover, bacterio-phage has the natural ability to infect bacteria, so that the efficiency of transfection is 100 times greater than it is for plasmids.

The construction of a DNA library begins with the isolation of genomic DNA and its digestion with a restriction enzyme to pro-duce fragments of 1,000 to 10,000 bp. These fragments are ligated into lambda genomes, which are subjected to a packaging reaction to produce mature viral particles, most of which carry a different piece of the genomic DNA. This collection of viruses is called a ge-nomic library and is used to study the structure and organization of specific genes. Clones from a library such as this contain the cod-ing sequences, in addition to noncoding sequences such as introns,

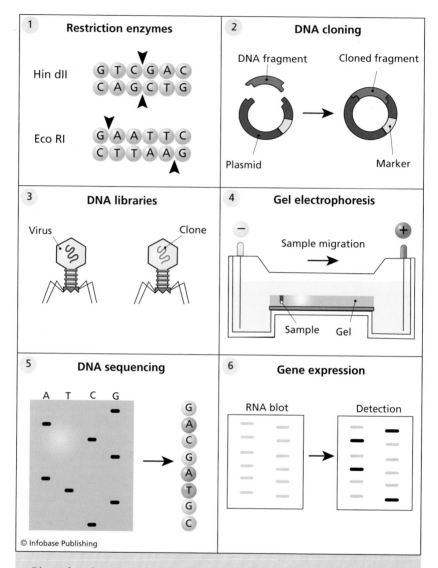

Biotechnology. This technology consists of six basic steps: 1) digestion of DNA with restriction enzymes in order to isolate specific DNA fragments; 2) cloning of restriction fragments in circular bacterial minichromosomes to increase their numbers; 3) storing the fragments for further study in viral-based DNA libraries; 4) isolation and purification of DNA fragments from gene libraries using gel electrophoresis; 5) sequencing cloned DNA fragments; and 6) determining the expression profile of selected DNA clones using RNA blots and radioactive detection procedures.

intervening sequences, promoters, and enhancers. An alternative form of a DNA library can be constructed by isolating mRNA from a specific cell type. This RNA is converted to the complementary DNA (cDNA) using an RNA-dependent DNA polymerase called reverse transcriptase. The cDNA is ligated to lambda genomes and packaged as for the genomic library. This collection of recombinant viruses is known as a cDNA library and contains genes that were being expressed by the cells when the mRNA was extracted. It does not include introns or controlling elements as these are lost during transcription and the processing that occurs in the cell to make mature mRNA. Thus, a cDNA library is intended for the purpose of studying gene expression and the structure of the coding region only.

Labeling Cloned DNA

Many of the procedures used in biotechnology were inspired by the events that occur during DNA replication (described above). This includes the labeling of cloned DNA for use as probes in expression studies, DNA sequencing, and polymerase chain reaction (PCR), which is described in a following section. DNA replication involves duplicating one of the strands (the parent, or template strand) by linking nucleotides in an order specified by the template and depends on a large number of enzymes, the most important of which is DNA polymerase. This enzyme, guided by the template strand, constructs a daughter strand by linking nucleotides together. One such nucleotide is deoxyadenine triphosphate (dATP). Deoxyribonucleotides have a single hydroxyl group located at the 3' carbon of the sugar group while the triphosphate is attached to the 5' carbon.

The procedure for labeling DNA probes, developed in 1983, introduces radioactive nucleotides into a DNA molecule. This method supplies DNA polymerase with a single-stranded DNA template, a primer, and the four nucleotides, in a buffered solution to induce in vitro replication. The daughter strand, which becomes the labeled probe, is made radioactive by including a ^{32}P-labeled nucleotide in the reaction mix. The radioactive nucleotide is usually deoxycytosine triphosphate (dCTP) or dATP. The ^{32}P is always part of the α (alpha) phosphate (the phosphate closest to the 5' carbon), as this is the one used by the polymerase to form the phosphodiester bond between nucleotides. Nucleotides can also be labeled with a fluorescent dye molecule.

Single-stranded DNA hexamers (six bases long) are used as primers, and these are produced in such a way that they contain all possible permutations of four bases taken six at a time. Randomizing the base sequence for the primers ensures that there will be at least one primer site in a template that is only 50 bp long. Templates used in labeling reactions such as this are generally 100 to 800 bp long. This strategy of labeling DNA is known as random primer labeling.

Gel Electrophoresis

This procedure is used to separate DNA and RNA fragments by size in a slab of agarose (highly refined agar) or polyacrylamide subjected to an electric field. Nucleic acids carry a negative charge and thus will migrate toward a positively charged electrode. The gel acts as a sieving medium that impedes the movement of the molecules. Thus, the rate at which the fragments migrate is a function of their size; small fragments migrate more rapidly than large fragments. The gel, containing the samples, is run submerged in a special pH-regulated solution, or buffer. Agarose gels are run horizontally as shown in panel 4 of the figure on page 188. But DNA sequencing gels, made of polyacrylamide, are much bigger and are run in a vertical tank.

DNA Sequencing

A sequencing reaction developed by the British biochemist Dr. Fred Sanger in 1976 is a technique that takes its inspiration from the natural process of DNA replication. DNA polymerase requires a primer with a free 3' hydroxyl group. The polymerase adds the first nucleotide to this group; and all subsequent bases are added to the 3' hydroxyl of the previous base. Sequencing by the Sanger method is usually performed with the DNA cloned into a special sequencing plasmid. This simplifies the choice of the primers since their sequence can be derived from the known plasmid sequence. Once the primer binds to the primer site the cloned DNA may be replicated.

Sanger's innovation involved the synthesis of chain-terminating nucleotide analogues lacking the 3' hydroxyl group. These analogues, also known as dideoxynucleotides (ddATP, ddCTP, ddGTP, and ddTTP), terminate the growth of the daughter strand at the point of insertion, and this can be used to determine the distance of each base on the daughter strand from the primer. These distances can be visualized by separating the Sanger reaction products on a polyacrylamide gel and then exposing the gel to X-ray film to produce an autoradiogram. The DNA sequence is read directly from this film beginning with the smallest fragment at the bottom of the gel (the nucleotide closest to the primer) and ending with the largest fragment at the top. A hypothetical autoradiogram and the derived DNA sequence are shown in panel 5 of the figure on page 188. The smallest fragment in this example is the "C" nucleotide at the bottom of lane 3. The next nucleotide in the sequence is the "G" nucleotide in lane 4, then the "T" nucleotide in lane 2, and so on to the top of the gel.

Automated versions of the Sanger sequencing reaction use fluorescent-labeled dideoxynucleotides, each with a different color, so the sequence of the template can be recorded by a computer as the reaction mix passes a sensitive photocell. Machines such as this were used to sequence the human genome—a job that cost many millions of dollars and took years to complete. Recent advances in DNA-sequencing technology will make it possible to sequence the human genome in less than a week at a cost of $1,000.

Gene Expression

The production of a genomic or cDNA library, followed by the sequencing of isolated clones, is a very powerful method for characterizing genes and the genomes from which they came. But the icing on the cake is the ability to determine the expression profile for a gene—that is, to determine which cells express the gene and exactly when the gene is turned on and off. Typical experiments may determine the expression of specific genes in normal versus cancerous tissue or tissues obtained from groups of different ages. There are essentially three methods for doing this: RNA blotting, fluorescent in situ hybridization (FISH), and PCR.

RNA Blotting

The procedure of RNA blotting consists of the following steps:

1. Extract mRNA from the cells or tissue of interest.
2. Fractionate (separate by size) the mRNA sample using gel electrophoresis.
3. Transfer the fractionated sample to a nylon membrane (the blotting step).
4. Incubate the membrane with a gene fragment (usually a cDNA clone) that has been labeled with a radioisotope.
5. Expose the membrane to X-ray film to visualize the signal.

The RNA is transferred from the gel to a nylon membrane using a vacuum apparatus or a simple dish containing a transfer buffer topped by a large stack of ordinary paper towels and a weight. The paper towels pull the transfer buffer through the gel, eluting the RNA from the gel and trapping it on the membrane. The location of specific mRNAs can be determined by hybridizing the membrane to a radiolabeled cDNA or genomic clone. The hybridization procedure involves placing the membrane in a buffer solution containing a labeled probe. During a long incubation period, the probe binds to the target sequence immobilized on the membrane. A-T and G-C base pairing (also known as hybridization) mediate the binding between the probe and target. The double-stranded molecule that is formed is a hybrid, being formed between the RNA target, on the membrane, and the DNA probe.

Fluorescent in Situ Hybridization (FISH)

Studying gene expression does not always depend on RNA blots and membrane hybridization. In the 1980s, scientists found that cDNA probes could be hybridized to DNA or RNA in situ, that is, while located within cells or tissue sections fixed on a microscope slide. In this case, the probe is labeled with a fluorescent dye molecule, rather than a radioactive isotope. The samples are then examined and photographed under a fluorescent microscope. FISH is an extremely powerful variation on RNA blotting. This procedure gives precise information regarding the identity of a cell that expresses a specific gene, information that usually cannot be obtained with membrane hybridization. Organs and tissues are generally composed of many different kinds of cells, which cannot be separated from one another using standard biochemical extraction procedures. Histological sections, however, show clearly the various cell types and, when subjected to FISH analysis, provide clear information as to which cells express specific genes. FISH is also used in clinical laboratories for the diagnosis of genetic abnormalities.

Polymerase Chain Reaction

PCR is simply repetitive DNA replication over a limited, primer-defined region of a suitable template. It provides a way of amplifying a short segment of DNA without going through the cloning procedures described above. The region defined by the primers is amplified to such an extent that it can be easily isolated for further study. The reaction exploits the fact that a DNA duplex, in a low salt buffer, will melt (i.e., separate into two single strands) at 167°F (75°C), but will re-anneal (rehybridize) at 98.6°F (37°C).

The reaction is initiated by melting the template in the presence of primers and polymerase in a suitable buffer, cooling quickly to 98.6°F (37°C), and allowing sufficient time for the polymerase to replicate both strands of the template. The temperature is then increased to 167°F (75°C) to melt the newly formed duplexes and then cooled to 98.6°F (37°C). At the lower temperature, more primer will anneal to initiate another round of replication. The heating-cooling cycle is repeated 20 to 30 times, after which the reaction products are fractionated on an agarose gel and the region containing the amplified fragment is cut out of the gel and purified for further study. The DNA polymerase used in these reactions is isolated from thermophilic bacteria that can withstand temperatures of 158°F (70°C) to 176°F (80°C). PCR applications are nearly limitless. It is used to amplify DNA from samples containing, at times, no more than a few cells. It is being used in the development of ultrafast DNA sequencers, identification of tissue samples in criminal investigations, amplification of ancient DNA obtained from fossils, and the identification of genes that are turned on or off during embryonic development or during cellular transformation (cancer formation).

GENE THERAPY

An illness is often due to invading microbes that destroy or damage cells and organs in our body. Cholera, smallpox, measles, diphtheria, AIDS, and the common cold are all examples of what is called an infectious disease. Such diseases may be treated with a drug that will, in some cases, remove the microbe from the body, thus curing the disease. Unfortunately, most diseases are not of the infectious kind. In such cases, there are no microbes to fight, no drugs to apply. Instead, physicians are faced with a far more difficult problem, for this type of disease is an ailment that damages a gene. Gene therapy attempts to cure these diseases by replacing, or supplementing, the damaged gene.

When a gene is damaged, it usually is caused by a point mutation, a change that affects a single nucleotide. Sickle-cell anemia, a disease affecting red blood cells, was the first genetic disorder of this kind to be described. The mutation occurs in a gene that codes for the β (beta) chain of hemoglobin, converting the codon GAG to GTG, which substitutes the amino acid valine at position 6 for glutamic acid. This single amino acid substitution is enough to cripple the hemoglobin molecule, making it impossible for it to carry enough oxygen to meet the demands of a normal adult. Scientists have identified several thousand genetic disorders that are known to be responsible for diseases such as breast cancer, colon cancer, hemophilia, and two neurological disorders, Alzheimer's disease and Parkinson's disease.

Gene therapy is made possible by recombinant DNA technology (biotechnology). Central to this technology is the use of viruses to clone specific pieces of DNA. That is, the DNA is inserted into a viral chromosome and is amplified as the virus multiplies. Viruses are parasites that specialize in infecting bacterial and animal cells. Consequently, scientists realized that a therapeutic gene could be inserted into a patient's cells by first introducing it into a virus and then letting the virus carry it into the affected cells. In this context the virus is referred to as a gene therapy delivery vehicle or vector (in recombinant technology it is referred to as a cloning vector).

Commonly used viruses are the retrovirus and the adenovirus. A retrovirus gets its name from the fact that it has an RNA genome that is copied into DNA after it infects a cell. Corona viruses (which cause the common cold) and the AIDS virus are common examples of retroviruses. The adenovirus (from "adenoid," a gland from which the virus was first isolated) normally infects the upper respiratory tract causing colds and flulike symptoms. This virus, unlike the retrovirus, has a DNA genome. Artificial vectors, called liposomes, have also been used that consist of a phospholipid vesicle (bubble) containing the therapeutic gene.

Gene therapy vectors are prepared by cutting the viral chromosome and the therapeutic gene with the same restriction enzyme, after which the two are joined together with a DNA ligase. This recombinant chromosome is packaged into viral particles to form the final vector. The vector may be introduced into cultured cells suffering from a genetic defect and then returned to the patient

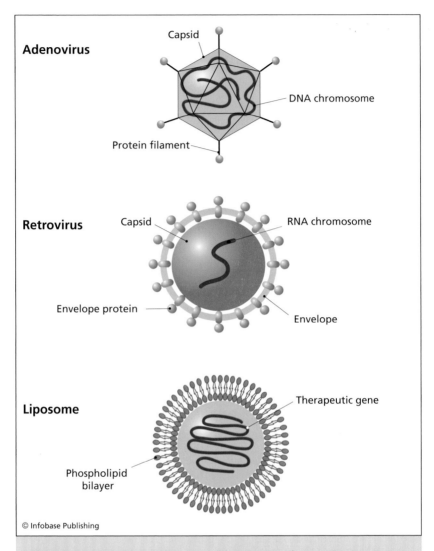

Adenovirus

Capsid

DNA chromosome

Protein filament

Retrovirus

Capsid

RNA chromosome

Envelope protein

Envelope

Liposome

Therapeutic gene

Phospholipid
bilayer

© Infobase Publishing

Vectors used in gene therapy. Adenoviruses have a DNA genome contained in a crystalline protein capsid and normally infect cells of the upper respiratory tract, causing colds and flulike symptoms. The protein filaments are used to infect cells. Retroviruses have an RNA genome that is converted to DNA when a cell is infected. The capsid is enclosed in a phospholipid envelope, studded with proteins that are used to infect cells. The AIDS virus is a common example of a retrovirus. Artificial vectors have also been used, consisting of a phospholipid bilayer enclosing the therapeutic gene.

from whom they were derived (ex vivo delivery). Alternatively, the vector may be injected directly into the patient's circulatory system (in vivo delivery). The ex vivo procedure is used when the genetic defect appears in white blood cells or stem cells that may be harvested from the patient and grown in culture. The in vivo procedure is used when the genetic defect appears in an organ, such as the liver, brain, or pancreas. This is the most common form of gene therapy, but it is also potentially hazardous because the vector, being free in the circulatory system, may infect a wide range of cells, thus activating an immune response that could lead to widespread tissue and organ damage.

The first gene therapy trial, conducted in 1990, used ex vivo delivery. This trial cured a young patient named Ashi deSilva of an immune deficiency (adenosine deaminase deficiency) that affects white blood cells. Other trials since then have either been ineffective or were devastating failures. Such a case occurred in 1999 when Jesse Gelsinger, an 18-year-old patient suffering from a liver disease, died while participating in a gene therapy trial. His death was caused by multiorgan failure brought on by the viral vector. In 2002, two children being treated for another form of immune deficiency developed vector-induced leukemia (cancer of the white blood cells). Subsequent studies, concluded in 2009, appear to have resolved these problems. Gene therapy holds great promise as a medical therapy. In the United States alone, there are currently more than 600 trials in progress to treat a variety of genetic disorders.

THE BELMONT REPORT

On July 12, 1975, the American National Research Act was signed into law, thereby creating a national commission to protect human research subjects. This commission was charged with the task of identifying basic ethical principles that should govern the conduct of any research involving human subjects, and in February 1976, the commission produced the Belmont Report (so named because the report was finalized at the Smithsonian Institution's Belmont

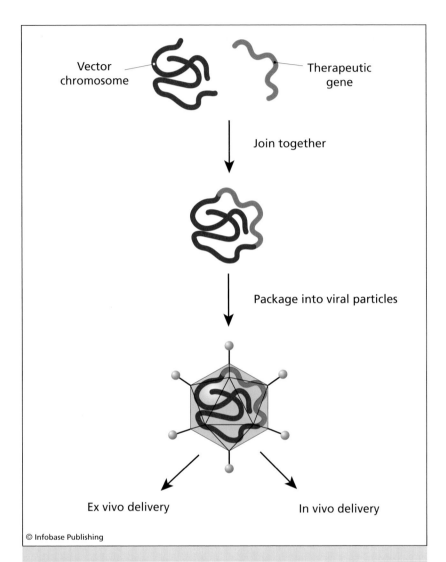

Vector chromosome

Therapeutic gene

Join together

Package into viral particles

Ex vivo delivery

In vivo delivery

© Infobase Publishing

Vector preparation and delivery. A viral chromosome and a therapeutic gene are cut with the same restriction enzyme and joined
together, after which the recombinant chromosome is packaged into
viral particles to form the vector. The vector may be introduced into
cultured cells and then returned to the patient from whom they were
derived (ex vivo delivery), or the vector may be injected directly into
the patient's circulatory system (in vivo delivery).

Conference Center). The report began by defining three basic ethical principles that should be applied to research involving human subjects: the principles of respect for persons, beneficence, and justice.

Respect for Persons

Respect for persons demands that subjects enter into research voluntarily and with adequate information. This assumes the individuals are autonomous agents, that is, are competent to make up their own minds. However, there are many instances where potential research subjects are not really autonomous: prisoners, patients in a mental institution, children, the elderly, and the infirm. All of these people require special protection to ensure they are not being coerced or fooled into volunteering as research subjects. The subjects in the Tuskegee study were all poor, uneducated farmworkers who were especially vulnerable to coercion.

Beneficence

Beneficence is generally regarded as acts of kindness or charity, but the report insisted that in the case of research subjects it be made an obligation. In this sense, it is the natural extension of the Hippocratic oath that all physicians are expected to adhere by: *I will give no deadly medicine to anyone if asked, nor suggest any such counsel.* In other words, physicians should do no harm; and those involved in biomedical research should never injure one person to benefit another.

Justice

Justice is an extension of beneficence. Researchers must never enlist subjects in an experiment if those subjects do not stand to reap any benefits. The exploitation of prisoners in Nazi concentration camps benefited the Nazis, but certainly not the people they experimented on. A second example cited by the Belmont Report was the Tuskegee study. Aside from committing a gross deviation from the most basic of ethical standards, the designers of the Tuskegee study enlisted only black people even though they are not the only racial group to suffer from this disease. The principle of justice was clearly not applied to these subjects.

Guided by these three ethical principles, the report introduced the following requirements that all human research trials must adhere to: informed consent, risk/benefit assessment, and fair selection of research subjects.

Informed Consent

All participants must provide informed consent in writing. Moreover, steps must be taken to ensure the consent is in fact informed. This might involve an independent assessment of the individual's ability to understand the language on the consent form and any instructions or explanations the investigators have given. Since the Gelsinger investigation, this process was amended to include a patient advocate, present at any meeting between the physicians and the prospective volunteers. This has the added advantage of ensuring that in a case where the patient is fully competent, the scientists do not give them misleading or inaccurate information or try to coerce them in any way.

Risk/Benefit Assessment

There is no point in having an ethical standard based on doing no harm if there is no formalized method available for assessing the risk to patient. It is the risk that is paramount in a patient's mind. No matter how grand the possible benefits, few would volunteer if they thought they would die as a consequence. The only exception to this might be terminally ill patients who volunteer for a clinical trial even though they know they are not likely to survive it. In general, risks should be reduced to those necessary to achieve the research objective. Risk assessment must be monitored by independent committees based on information supplied by the investigators. If there is significant risk, review committees are expected to demand a justification for it.

Selection of Subjects

The selection process must be fair. Low-risk, potentially beneficial research should not be offered to one segment of our society, while high-risk research is conducted on prisoners, low-income groups, or anyone in a disadvantaged social position.

Conclusions

The Belmont Report introduced, for the first time, the principle of informed consent. Backing this up are independent review committees that ensure the ethical guidelines are being followed. In the United States, the Food and Drug Administration (FDA) and the National Institutes of Health (NIH) are responsible for enforcing the guidelines laid out by the Belmont Report. There are, in addition, local review committees (called institutional review boards) that must approve any experimentation using human subjects. The Belmont Report was inspired by the general public's anger over the Tuskegee study. It is fitting that on May 16, 1997, the surviving members of the Tuskegee study were invited to the White House, where President Bill Clinton issued a formal apology and reaffirmed the nation's commitment to rigorous ethical standards in biomedical research. No one would have believed at the time that further trouble was just around the corner.

THE GELSINGER INVESTIGATION

In fall 1998, a gene therapy trial to treat a liver disease was begun at the University of Pennsylvania. The investigators recruited 18 patients, and the 18th patient, who happened to be 18 years of age, was Jesse Gelsinger. Gelsinger joined the trial on September 13, 1999. On the second day of his treatment, he lapsed into a coma and was pronounced dead 24 hours later. Within days of Gelsinger's death, the NIH ordered a halt to all American gene therapy trials using a similar research protocol. The ban was to last a full year and was accompanied by an investigation that was not concluded until fall 2001.

The trial team leader, Dr. James Wilson, reported Gelsinger's death immediately. A preliminary review was conducted from November 30, 1999, to January 19, 2000. The full review was to last for more than a year and covered every aspect of Dr. Wilson's protocol and the criteria used to admit patients to the trial. In January 2000, the NIH released the preliminary results of their investigation, which cited the principal investigators for failure to adhere to the clinical protocol and an apparent disregard for the safety of the study subjects. The report focused on four main points: failure to adhere to the stopping rules, failure to adhere to the principle of informed consent, failure to keep adequate records regarding vector lineage and titer, and changing the protocol without approval.

Failure to Adhere to the Stopping Rules

The study was designed around several cohorts that were treated in tandem so that in the event of toxic reactions in one cohort treatment the study could be terminated before other cohorts were treated. However, toxic reactions observed in five of the cohorts did not lead to termination of the trial before Gelsinger was treated. Many of the patients suffered harsher reactions to the treatment than was expected, and this should have been sufficient reason to stop the trial. In addition, most of the toxic reactions experienced by the patients in this study were never reported to the FDA or the NIH. In the months following the conclusion of the Gelsinger preliminary investigation, other investigations showed that failure to report toxic reactions was a common failure in many gene therapy trials. In one study, the patients experienced 691 serious side effects and, of these, only 39 were reported as required by the federal agencies.

Failure to Adhere to the Principle of Informed Consent

When a toxic response occurred in cohort one, cohort two should have been informed of this response to give those patients the option of withdrawing from the study. This was not done. Moreover, the investigators discovered that none of the subjects were told about adverse effects of monkeys in the preclinical trial. One of the monkeys received the same virus used in the clinical trial, though at a higher dose, and within a week of being treated it was euthanized because it developed the same clotting disorder that killed Gelsinger. Since the subjects were not told about this, the consent forms were ruled invalid. It was this charge that led to the call for a patient advocate in all future biomedical research trials, regardless of their nature.

Failure to Keep Adequate Records Regarding Vector Lineage and Titer

This was an especially damaging finding, since it implied that the researchers gave Gelsinger more virus than they thought they had. The term *titer* refers to the number of vector particles in a given solution. Determining the titer is not straightforward, and, if errors are made, the concentration may be out in increments of 10, rather than double or triple the amount expected. The possibility that Gelsinger was accidentally given a higher-than-stated dose is suggested by the fact that a woman in his cohort received a nearly identical dose (3.0×10^{13}) without signs of liver damage or multiorgan failure. As mentioned above, a monkey in a preclinical trial received a higher dose (17× greater) of the same virus and subsequently died of multiorgan failure. If there was an error made in calculating the dose for Gelsinger, it is possible he received an equivalent, fatal amount.

Changing the Protocol without Approval

The most serious infraction here had to do with the ammonia levels in the blood of prospective volunteers. As laid out in the original protocol, patients having more that 50 micromoles of ammonia per milliliter of blood were barred from volunteering, because such a test result indicates severe liver damage. This was increased, sometime after the trial began, to 70 micromoles, without formal approval from the FDA. Gelsinger's ammonia level, on the day he was treated, was about 60 micromoles. If the original cutoff had been adhered to, he would have been excluded from the study. This is another indication of how important it is to adhere to the principle of informed consent and to the inclusion of an independent patient advocate.

UNDERSTANDING CLINICAL TRIALS

Clinical trials are conducted in four phases and are always preceded by research conducted on experimental animals such as mice, rats, or monkeys. The format for preclinical research is informal. It is conducted in a variety of research labs around the world, and the results are published in scientific journals. Formal approval from a governmental regulatory body is not required.

Phase I Clinical Trial

Pending the outcome of the preclinical research, investigators may apply for permission to try the experiments on human subjects. Applications in the United States are made to the FDA, the NIH, and the Recombinant DNA Advisory Committee (RAC). The RAC was set up by the NIH to monitor any research, including clinical trials, dealing with cloning, recombinant DNA, or gene therapy. Phase I trials are conducted on a small number of adult volunteers, usually between two and 20, who have given informed consent. That is, the investigators explain the procedure, the possible outcomes, and, especially, the dangers associated with the procedure before the subjects sign a consent form. The purpose of the Phase I trial is to determine the overall effect the treatment has on humans. A treatment that works well in monkeys or mice may not work at all on humans. Similarly, a treatment that appears safe in lab animals may be toxic, even deadly, when given to humans. Since most clinical trials are testing a new drug of some kind, the first priority is to determine a safe dosage for humans. Consequently, subjects in the Phase I trial are given a range of doses, all of which, even the highest, are less than the highest dose given to experimental animals. If the results from the Phase I trial are promising, the investigators may apply for permission to proceed to Phase II.

Phase II Clinical Trial

Having established the general protocol, or procedure, the investigators now try to replicate the encouraging results from Phase I, but with a much larger number of subjects (100–300). Only with a large number of subjects is it possible to prove the treatment has an effect. In addition, dangerous side effects may have been missed in Phase I because of a small sample size. The results from Phase II will determine how safe the procedure is and whether it works or not. If the statistics show the treatment is effective and toxicity is low, the investigators may apply for permission to proceed to Phase III.

Phase III Clinical Trial

Based on Phase II results, the procedure may look very promising, but before it can be used as a routine treatment it must be tested on thousands of patients at a variety of research centers. This is the expensive part of bringing a new drug or therapy to market, costing millions, sometimes billions, of dollars. It is for this reason that Phase III clinical trials invariably have the financial backing of large pharmaceutical or biotechnology companies. If the results of the Phase II trial are confirmed in Phase III, the FDA will approve the use of the drug for routine treatment. The use of the drug or treatment now passes into an informal Phase IV trial.

Phase IV Clinical Trial

Even though the treatment has gained formal approval, its performance is monitored for very long-term effects, sometimes stretching on for 10 to 20 years. In this way, the FDA retains the power to recall the drug long after it has become a part of standard medical procedure. It can happen that in the long term, the drug costs more than an alternative, in which case, health insurance providers may refuse to cover the cost of the treatment.

GENE AND PROTEIN NOMENCLATURE

Scientists who were, in effect, probing around in the dark have discovered many genes and their encoded proteins. Once discovered, the new genes or proteins had to be named. Usually the name is nothing more than a lab book code or an acronym suggested by the system under study at the time. Sometimes it turns out, after further study, that the function observed in the original study is a minor aspect of the gene's role in the cell. It is for this reason that gene and protein names sometimes seem absurd and poorly chosen.

In 2003, an International Committee on Standardized Genetic Nomenclature agreed to unify the rules and guidelines for gene and protein names for the mouse and rat. Similar committees have attempted to standardize gene-naming conventions for human, frog, zebrafish, and yeast genes. In general, the gene name is expected to be brief and to begin with a lower case letter unless it is a person's name. The gene symbols are acronyms taken from the gene name and are expected to be three to five characters long and not more than 10. The symbols must be written with Roman letters and Arabic numbers. The same symbol is used for orthologs (i.e., the same gene) among different species, such as human, mouse, or rat. Thus, the gene sonic hedgehog is symbolized as shh and the gene myelocytomatosis is symbolized as myc.

Unfortunately, the various committees were unable to agree on a common presentation for the gene and protein symbols. A human gene symbol, for example, is italicized with uppercase letters, and the protein is uppercase and not italicized. A frog gene symbol is lowercase, and the protein is uppercase, while neither is italicized. Thus, the myc gene and its protein, for example, are written as *MYC* and MYC in humans, myc and MYC in frogs, and *Myc* and Myc in mice and rats. The latter convention, *Myc* and Myc is used throughout this set, regardless of the species.

WEIGHTS AND MEASURES

The following table presents some common weights, measures, and conversions that appear in this book.

QUANTITY	EQUIVALENT
length	1 meter (m) = 100 centimeters (cm) = 1.094 yards = 39.37 inches 1 kilometer (km) = 1,000 m = 0.62 miles 1 foot = 30.48 cm 1 inch = 1/12 foot = 2.54 cm 1 cm = 0.394 inch = 10^{-2} (or 0.01) m 1 millimeter (mm) = 10^{-3} m 1 micrometer (µm) = 10^{-6} m 1 nanometer (nm) = 10^{-9} m 1 Ångström (Å) = 10^{-10} m
mass	1 gram (g) = 0.0035 ounce 1 pound = 16 ounces = 453.6 grams 1 kilogram (kg) = 2.2 pounds (lb) 1 milligram (mg) = 10^{-3} g 1 microgram (µg) = 10^{-6} g
volume	1 liter (l) = 1.06 quarts (US) = 0.264 gallon (US) 1 quart (US) = 32 fluid ounces = 0.95 liter 1 milliliter (ml) = 10^{-3} liter = 1 cubic centimeter (cc)
temperature	°C = 5/9 (°F - 32) °F = (9/5 × °C) + 32
energy	calorie = the amount of heat needed to raise the temperature of 1 gram of water by 1°C. kilocalorie = 1,000 calories. Used to describe the energy content of foods.

Glossary

acetyl   A chemical group derived from acetic acid that is important in energy metabolism and for the modification of proteins.

acetylcholine   A neurotransmitter released at axonal terminals by cholinergic neurons, found in the central and peripheral nervous systems and released at the vertebrate neuromuscular junction.

acetyl-CoA   A water-soluble molecule, coenzyme A (CoA) that carries acetyl groups in cells.

acid   A substance that releases protons when dissolved in water; carries a net negative charge.

actin filament   A protein filament formed by the polymerization of globular actin molecules; forms the cytoskeleton of all eukaryotes and part of the contractile apparatus of skeletal muscle.

action potential   A self-propagating electrical impulse that occurs in the membranes of neurons, muscles, photoreceptors, and hair cells of the inner ear.

active transport   Movement of molecules across the cell membrane, using the energy stored in ATP.

adenylate cyclase   A membrane-bound enzyme that catalyzes the conversion of ATP to cyclic AMP; an important component of cell-signaling pathways.

adherens junction   A cell junction in which the cytoplasmic face of the membrane is attached to actin filaments.

adipocyte   A fat cell.

adrenaline (epinephrine)   A hormone released by chromaffin cells in the adrenal gland; prepares an animal for extreme activity by increasing the heart rate and blood sugar levels.

adult stem cells   Stem cells isolated from adult tissues, such as bone marrow or epithelium.

aerobic   Refers to a process that either requires oxygen or occurs in its presence.

agar   A polysaccharide isolated from seaweed that forms a gel when boiled in water and cooled to room temperature; used by microbiologists as a solid culture medium for the isolation and growth of bacteria and fungi.

agarose   A purified form of agar that is used to fractionate (separate by size) biomolecules.

allele   An alternate form of a gene. Diploid organisms have two alleles for each gene, located at the same locus (position) on homologous chromosomes.

allogeneic transplant   A cell, tissue, or organ transplant from an unrelated individual.

alpha helix   A common folding pattern of proteins in which a linear sequence of amino acids twists into a right-handed helix stabilized by hydrogen bonds.

amino acid   An organic molecule containing amino and carboxyl groups that is a building block of protein.

aminoacyl tRNA   An amino acid linked by its carboxyl group to a hydroxyl group on tRNA.

aminoacyl-tRNA synthetase   An enzyme that attaches the correct amino acid to a tRNA.

amino terminus   The end of a protein or polypeptide chain that carries a free amino group.

amphipathic   Having both hydrophilic and hydrophobic regions, as in a phospholipid.

anabolism   A collection of metabolic reactions in a cell whereby large molecules are made from smaller ones.

anaerobic   A cellular metabolism that does not depend on molecular oxygen.

anaphase   A mitotic stage in which the two sets of chromosomes move away from each other toward opposite spindle poles.

anchoring junction   A cell junction that attaches cells to each other.

angiogenesis   Sprouting of new blood vessels from preexisting ones.

angstrom   A unit of length, equal to 10^{-10} meter or 0.1 nanometer (nM), that is used to measure molecules and atoms.

anterior   A position close to or at the head end of the body.

antibiotic   A substance made by bacteria, fungi, and plants that is toxic to microorganisms. Common examples are penicillin and streptomycin.

antibody   A protein made by B cells of the immune system in response to invading microbes.

anticodon   A sequence of three nucleotides in tRNA that is complementary to a messenger RNA codon.

antigen   A molecule that stimulates an immune response, leading to the formation of antibodies.

antigen-presenting cell   A cell of the immune system, such as a monocyte, that presents pieces of an invading microbe (the antigen) to lymphocytes.

antiparallel   The relative orientation of the two strands in a DNA double helix; the polarity of one strand is oriented in the opposite direction to the other.

antiporter   A membrane carrier protein that transports two different molecules across a membrane in opposite directions.

apoptosis   Regulated or programmed form of cell death that may be activated by the cell itself or by the immune system to force cells to commit suicide when they become infected with a virus or bacterium.

archaea   The archaea are prokaryotes that are physically similar to bacteria (both lack a nucleus and internal organelles), but they have retained a primitive biochemistry and physiology that would have been commonplace 2 billion years ago.

asexual reproduction   The process of forming new individuals without gametes or the fertilization of an egg by a sperm. Individuals produced this way are identical to the parent and referred to as a clone.

aster   The star-shaped arrangement of microtubules that is characteristic of a mitotic or meiotic spindle.

ATP (adenosine triphosphate)   A nucleoside consisting of adenine, ribose, and three phosphate groups that is the main carrier of chemical energy in the cell.

ATPase   Any enzyme that catalyzes a biochemical reaction by extracting the necessary energy from ATP.

ATP synthase   A protein located in the inner membrane of the mitochondrion that catalyzes the formation of ATP from ADP and inorganic phosphate using the energy supplied by the electron transport chain.

autogeneic transplant   A patient receives a transplant of his or her own tissue.

autologous   Refers to tissues or cells derived from the patient's own body.

autoradiograph (autoradiogram)   X-ray film that has been exposed to X-rays or to a source of radioactivity; used to visualize internal structures of the body and radioactive signals from sequencing gels and DNA or RNA blots.

autosome   Any chromosome other than a sex chromosome.

axon   A long extension of a neuron's cell body that transmits an electrical signal to other neurons.

axonal transport   The transport of organelles, such as Golgi vesicles, along an axon to the axonal terminus. Transport also flows from the terminus to the cell body.

bacteria   One of the most ancient forms of cellular life (the other is the archaea). Bacteria are prokaryotes, and some are known to cause disease.

bacterial artificial chromosome (BAC)   A cloning vector that accommodates DNA inserts of up to 1 million base pairs.

bacteriophage   A virus that infects bacteria. Bacteriophages were used to prove that DNA is the cell's genetic material and are now used as cloning vectors.

base   A substance that can accept a proton in solution. The purines and pyrimidines in DNA and RNA are organic bases and are often referred to simply as bases.

base pair   Two nucleotides in RNA or DNA that are held together by hydrogen bonds. Adenine bound to thymine or guanine bound to cytosine are examples of base pairs

B cell (B lymphocyte)   A white blood cell that makes antibodies and is part of the adaptive immune response.

benign   Tumors that grow to a limited size and do not spread to other parts of the body.

beta sheet   Common structural motif in proteins in which different strands of the protein run alongside one another and are held together by hydrogen bonds.

biopsy   The removal of cells or tissues for examination under a microscope. When only a sample of tissue is removed, the procedure is called an incisional biopsy or core biopsy. When an entire lump or suspicious area is removed, the procedure is called an excisional biopsy. When a sample of tissue or fluid is removed with a needle, the procedure is called a needle biopsy or fine-needle aspiration.

biosphere   The world of living organisms

biotechnology   A set of procedures that are used to study and manipulate genes and their products.

blastomere   A cell formed by the cleavage of a fertilized egg. Blastomeres are the totipotent cells of the early embryo.

blotting   A technique for transferring DNA (southern blotting), RNA (northern blotting), or proteins (western blotting) from an agarose or polyacrylamide gel to a nylon membrane.

BRCA1 (breast cancer gene 1)   A gene on chromosome 17 that may be involved in regulating the cell cycle. A person who inherits an altered version of the BRCA1 gene has a higher risk of getting breast, ovarian, or prostate cancer.

BRCA2 (breast cancer gene 2)   A gene on chromosome 13 that, when mutated, increases the risk of getting breast, ovarian, or prostate cancer.

budding yeast   The common name for the baker's yeast *Saccharomyces cerevisiae*, a popular experimental organism that reproduces by budding off a parental cell.

buffer   A pH-regulated solution with a known electrolyte (salt) content; used in the isolation, manipulation, and storage of biomolecules and medicinal products.

cadherin   Belongs to a family of proteins that mediates cell-cell adhesion in animal tissues.

calorie   A unit of heat. One calorie is the amount of heat needed to raise the temperature of one gram of water by 1°C. kilocalories (1,000 calories) are used to describe the energy content of foods.

capsid   The protein coat of a virus, formed by autoassembly of one or more proteins into a geometrically symmetrical structure.

carbohydrate   A general class of compounds that includes sugars, containing carbon, hydrogen, and oxygen.

carboxyl group   A carbon atom attached to an oxygen and a hydroxyl group

carboxyl terminus   The end of a protein containing a carboxyl group.

carcinogen   A compound or form of radiation that can cause cancer.

carcinogenesis   The formation of a cancer.

carcinoma   Cancer of the epithelium, representing the majority of human cancers.

cardiac muscle   Muscle of the heart; composed of myocytes that are linked together in a communication network based on free passage of small molecules through gap junctions.

caspase   A protease involved in the initiation of apoptosis.

catabolism   Enzyme regulated breakdown of large molecules for the extraction of chemical-bond energy. Intermediate products are called catabolites.

catalyst   A substance that lowers the activation energy of a reaction.

CD28   Cell-surface protein located in T-cell membranes, necessary for the activation of T-cells by foreign antigens.

cDNA (complementary DNA)   DNA that is synthesized from mRNA, thus containing the complementary sequence; cDNA contains coding sequence, but not the regulatory sequences that are present in the genome. Labeled probes are made from cDNA for the study of gene expression.

cell adhesion molecule (CAM)   A cell surface protein that is used to connect cells to one another.

cell body   The main part of a cell containing the nucleus, Golgi complex, and endoplasmic reticulum; used in reference to neurons that have long processes (dendrites and axons) extending some distance from the nucleus and cytoplasmic machinery.

cell coat   (see **glycocalyx**)

cell-cycle control system   A team of regulatory proteins that governs progression through the cell cycle.

cell-division-cycle gene (*cdc* gene)   A gene that controls a specific step in the cell cycle.

cell fate   The final differentiated state that a pluripotent embryonic cell is expected to attain.

cell-medicated immune response   Activation of specific cells to launch an immune response against an invading microbe.

cell nuclear transfer   Animal cloning technique whereby a somatic cell nucleus is transferred to an enucleated oocyte; synonymous with somatic cell nuclear transfer.

celsius   A measure of temperature. This scale is defined such that 0°C is the temperature at which water freezes and 100°C is the temperature at which water boils.

central nervous system (CNS)   That part of a nervous system that analyzes signals from the body and the environment. In animals, the CNS includes the brain and spinal cord.

centriole   A cylindrical array of microtubules that is found at the center of a centrosome in animal cells.

centromere   A region of a mitotic chromosome that holds sister chromatids together. Microtubules of the spindle fiber connect to an area of the centromere called the kinetochore.

centrosome   Organizes the mitotic spindle and the spindle poles; in most animal cells it contains a pair of centrioles.

chiasma (plural chiasmata)   An X-shaped connection between homologous chromosomes that occurs during meiosis I, representing a site of crossing-over, or genetic exchange between the two chromosomes.

chromatid   A duplicate chromosome that is still connected to the original at the centromere. The identical pair are called sister chromatids.

chromatin   A complex of DNA and proteins (histones and non-histones) that forms each chromosome and is found in the nucleus of all eukaryotes. Decondensed and threadlike during interphase.

chromatin condensation   Compaction of different regions of interphase chromosomes that is mediated by the histones.

chromosome   One long molecule of DNA that contains the organism's genes. In prokaryotes, the chromosome is circular and naked; in eukaryotes, it is linear and complexed with histone and nonhistone proteins.

chromosome condensation   Compaction of entire chromosomes in preparation for cell division.

clinical breast exam   An exam of the breast performed by a physician to check for lumps or other changes.

cnidoblast   A stinging cell found in the Cnidarians (jellyfish).

cyclic adenosine monophosphate (cAMP)   A second messenger in a cell-signaling pathway that is produced from ATP by the enzyme adenylate cyclase.

cyclin   A protein that activates protein kinases (cyclin-dependent protein kinases, or Cdk) that control progression from one stage of the cell cycle to another.

cytochemistry   The study of the intracellular distribution of chemicals.

cytochrome   Colored, iron-containing protein that is part of the electron transport chain.

cytotoxic T cell   A T lymphocyte that kills infected body cells.

dendrite   An extension of a nerve cell that receives signals from other neurons.

dexrazoxane   A drug used to protect the heart from the toxic effects of anthracycline drugs such as doxorubicin. It belongs to the family of drugs called chemoprotective agents.

dideoxynucleotide   A nucleotide lacking the 2' and 3' hydroxyl groups.

dideoxy sequencing   A method for sequencing DNA that employs dideoxyribose nucleotides; also known as the Sanger sequencing method, after Fred Sanger, a chemist who invented the procedure in 1976.

diploid   A genetic term meaning two sets of homologous chromosomes, one set from the mother and the other from the father. Thus, diploid organisms have two versions (alleles) of each gene in the genome.

DNA (deoxyribonucleic acid)   A long polymer formed by linking four different kinds of nucleotides together likes beads on a string. The sequence of nucleotides is used to encode an organism's genes.

DNA helicase   An enzyme that separates and unwinds the two DNA strands in preparation for replication or transcription.

DNA library   A collection of DNA fragments that are cloned into plasmids or viral genomes.

DNA ligase   An enzyme that joins two DNA strands together to make a continuous DNA molecule.

DNA microarray   A technique for studying the simultaneous expression of a very large number of genes.

DNA polymerase   An enzyme that synthesizes DNA using one strand as a template.

DNA primase   An enzyme that synthesizes a short strand of RNA that serves as a primer for DNA replication.

dorsal   The backside of an animal; also refers to the upper surface of anatomical structures, such as arms or wings.

dorsalventral   The body axis running from the backside to the frontside or the upperside to the underside of a structure.

double helix   The three-dimensional structure of DNA in which the two strands twist around each other to form a spiral.

doxorubicin   An anticancer drug that belongs to a family of antitumor antibiotics.

Drosophila melanogaster   Small species of fly, commonly called a fruit fly, that is used as an experimental organism in genetics, embryology, and gerontology.

ductal carcinoma in situ (DCIS)   Abnormal cells that involve only the lining of a breast duct. The cells have not spread outside the duct to other tissues in the breast; also called intraductal carcinoma.

dynein   A motor protein that is involved in chromosome movements during cell division.

dysplasia   Disordered growth of cells in a tissue or organ, often leading to the development of cancer.

ectoderm   An embryonic tissue that is the precursor of the epidermis and the nervous system.

electrochemical gradient   A differential concentration of an ion or molecule across the cell membrane that serves as a source of potential energy and may polarize the cell electrically.

electron microscope   A microscope that uses electrons to produce a high-resolution image of the cell.

electrophoresis   The movement of a molecule, such as protein, DNA, or RNA, through an electric field. In practice, the molecules migrate through a slab of agarose or polyacrylamide that is immersed in a special solution and subjected to an electric field.

elution To remove one substance from another by washing it out with a buffer or solvent.

embryogenesis The development of an embryo from a fertilized egg.

embryonic stem cell (ES cell) A pluripotent cell derived from the inner cell mass (the cells that give rise to the embryo instead of the placenta) of a mammalian embryo.

endocrine cell A cell that is specialized for the production and release of hormones. Such cells make up hormone-producing tissue such as the pituitary gland or gonads.

endocytosis Cellular uptake of material from the environment by invagination of the cell membrane to form a vesicle called an endosome. The endosome's contents are made available to the cell after it fuses with a lysosome.

endoderm An embryonic tissue layer that gives rise to the gut.

endoplasmic reticulum (ER) Membrane-bounded chambers that are used to modify newly synthesized proteins with the addition of sugar molecules (glycosylation). When finished, the glycosylated proteins are sent to the Golgi apparatus in exocytotic vesicles.

enhancer A DNA-regulatory sequence that provides a binding site for transcription factors capable of increasing the rate of transcription for a specific gene; often located thousands of base pairs away from the gene it regulates.

enveloped virus A virus containing a capsid that is surrounded by a lipid bilayer originally obtained from the membrane of a previously infected cell.

enzyme A protein or RNA that catalyzes a specific chemical reaction.

epidermis The epithelial layer, or skin, that covers the outer surface of the body.

ER marker sequence The amino terminal sequence that directs proteins to enter the endoplasmic reticulum (ER). This sequence is removed once the protein enters the ER.

erythrocyte A red blood cell that contains the oxygen-carrying pigment hemoglobin; used to deliver oxygen to cells in the body.

Escherichia coli **(E. coli)** Rod-shape, gram-negative bacterium that inhabits the intestinal tract of most animals and is used as an experimental organism by geneticists and biomedical researchers.

euchromatin Lightly staining portion of interphase chromatin, in contrast to the darkly staining heterochromatin (condensed chromatin). Euchromatin contains most, if not all, of the active genes.

eukaryote (eucaryote) A cell containing a nucleus and many membrane-bounded organelles. All life-forms, except bacteria and viruses, are composed of eukaryote cells.

exocytosis The process by which molecules are secreted from a cell. Molecules to be secreted are located in Golgi-derived vesicles that fuse with the inner surface of the cell membrane, depositing the contents into the intercellular space.

exon Coding region of a eukaryote gene that is represented in messenger RNA and thus directs the synthesis of a specific protein.

expression studies Examination of the type and quantity of mRNA or protein that is produced by cells, tissues, or organs.

fat A lipid material, consisting of triglycerides (fatty acids bound to glycerol), that is stored adipocytes as an energy reserve.

fatty acid A compound that has a carboxylic acid attached to a long hydrocarbon chain. A major source of cellular energy and a component of phospholipids.

fertilization The fusion of haploid male and female gametes to form a diploid zygote.

fibroblast The cell type that, by secreting an extracellular matrix, gives rise to the connective tissue of the body.

Filopodium A fingerlike projection of a cell's cytoplasmic membrane, commonly observed in amoeba and embryonic nerve cells.

filter hybridization The detection of specific DNA or RNA molecules, fixed on a nylon filter (or membrane), by incubating the filter with a labeled probe that hybridizes to the target sequence; also known as membrane hybridization.

fixative A chemical that is used to preserve cells and tissues. Common examples are formaldehyde, methanol, and acetic acid.

flagellum (plural flagella) Whiplike structure found in prokaryotes and eukaryotes that are used to propel cells through water.

fluorescein Fluorescent dye that produces a green light when illuminated with ultraviolet or blue light.

fluorescent dye A dye that absorbs UV or blue light and emits light of a longer wavelength, usually as green or red light.

fluorescent in situ hybridization (FISH) A procedure for detecting the expression of a specific gene in tissue sections or smears through the use of DNA probes labeled with a fluorescent dye.

fluorescent microscope A microscope that is equipped with special filters and a beam splitter for the examination of tissues and cells stained with a fluorescent dye.

follicle cell Cells that surround and help feed a developing oocyte.

G_0 G "zero" refers to a phase of the cell cycle; state of withdrawal from the cycle as the cell enters a resting or quiescent stage; occurs in differentiated body cells, as well as in developing oocytes.

G_1 Gap 1 refers to the phase of the cell cycle that occurs just after mitosis and before the next round of DNA synthesis.

G_2 The Gap 2 phase of the cell cycle follows DNA replication and precedes mitosis.

gap junction A communication channel in the membranes of adjacent cells that allows free passage of ions and small molecules.

gel electrophoresis A procedure that is used to separate biomolecules by forcing them to migrate through a gel matrix (agarose or polyacrylamide) subjected to an electric field.

gene A region of the DNA that specifies a specific protein or RNA molecule that is handed down from one generation to the next. This region includes both the coding, noncoding, and regulatory sequences.

gene regulatory protein Any protein that binds to DNA and thereby affects the expression of a specific gene.

gene repressor protein A protein that binds to DNA and blocks transcription of a specific gene.

gene therapy A method for treating disease whereby a defective gene, causing the disease, is either repaired, replaced, or supplemented with a functional copy.

genetic code A set of rules that assigns a specific DNA or RNA triplet, consisting of a three-base sequence, to a specific amino acid.

genome All of the genes that belong to a cell or an organism.

genomic library A collection of DNA fragments, obtained by digesting genomic DNA with a restriction enzyme, that are cloned into plasmid or viral vectors.

genomics The study of DNA sequences and their role in the function and structure of an organism.

genotype The genetic composition of a cell or organism.

germ cell Cells that develop into gametes, either sperm or oocytes.

glucose Six-carbon monosaccharide (sugar) that is the principal source of energy for many cells and organisms; stored as glycogen in animal cells and as starch in plants. Wood is an elaborate polymer of glucose and other sugars.

glycerol A three-carbon alcohol that is an important component of phospholipids.

glycocalyx A molecular "forest," consisting of glycosylated proteins and lipids, that covers the surface of every cell. The glycoproteins and glycolipids, carried to the cell membrane by Golgi-derived vesicles, have many functions including the formation of ion channels, cell-signaling receptors, and transporters.

glycogen A polymer of glucose, used to store energy in an animal cell.

glycolysis The degradation of glucose with production of ATP.

glycoprotein Any protein that has a chain of glucose molecules (oligosaccharide) attached to some of the amino acid residues.

glycosylation The process of adding one or more sugar molecules to proteins or lipids.

glycosyl transferase An enzyme in the Golgi complex that adds glucose to proteins.

Golgi complex (Golgi apparatus) Membrane-bounded organelle in eukaryote cells that receives glycoproteins from the ER, which are modified and sorted before being sent to their final destination. The Golgi complex is also the source of glycolipids that are destined for the cell membrane. The glycoproteins and glycolipids leave the Golgi by exocytosis. This organelle is named after the Italian histologist Camillo Golgi, who discovered it in 1898.

Gram stain A bacterial stain that detects different species of bacteria based on the composition of their cell wall. Bacteria that retain the Gram stain are colored blue (Gram positive), whereas those that do not are colored orange (Gram negative).

granulocyte A type of white blood cell that includes the neutrophils, basophils, and eosinophils.

growth factor   A small protein (polypeptide) that can stimulate cells to grow and proliferate.

haploid   Having only one set of chromosomes; a condition that is typical in gametes, such as sperm and eggs.

HeLa cell   A tumor-derived cell line, originally isolated from a cancer patient in 1951; currently used by many laboratories to study the cell biology of cancer and carcinogenesis.

helix-loop-helix   A structural motif common to a group of gene-regulatory proteins.

helper T cell   A type of T lymphocyte that helps stimulate B cells to make antibodies directed against a specific microbe or antigen.

hemoglobin   An iron-containing protein complex, located in red blood cells, that picks up oxygen in the lungs and carries it to other tissues and cells of the body.

hemopoiesis   Production of blood cells, occurring primarily in the bone marrow.

hematopoietic   Refers to cells, derived form the bone marrow, that give rise to red and white blood cells.

hematopoietic stem cell transplantation (HSCT)   The use of stem cells isolated from the bone marrow to treat leukemia and lymphoma.

hepatocyte   A liver cell.

heterochromatin   A region of a chromosome that is highly condensed and transcriptionally inactive.

histochemistry   The study of chemical differentiation of tissues.

histology   The study of tissues.

histone   Small nuclear proteins, rich in the amino acids arginine and lysine, that form the nucleosome in eukaryote nuclei, a beadlike structure that is a major component of chromatin.

HIV   The human immunodeficiency virus that is responsible for AIDS.

homolog   One of two or more genes that have a similar sequence and are descended from a common ancestor gene.

homologous   Organs or molecules that are similar in structure because they have descended from a common ancestor; used primarily in reference to DNA and protein sequences.

homologous chromosomes   Two copies of the same chromosome, one inherited from the mother and the other from the father.

hormone   A signaling molecule, produced and secreted by endocrine glands; usually released into general circulation for coordination of an animal's physiology.

housekeeping gene   A gene that codes for a protein that is needed by all cells, regardless of the cell's specialization. Genes encoding enzymes involved in glycolysis and Krebs cycle are common examples.

hybridization   A term used in molecular biology (recombinant DNA technology) meaning the formation a double-stranded nucleic acid through complementary base-pairing; a property that is exploited in filter hybridization; a procedure that is used to screen gene libraries and to study gene structure and expression.

hydrolysis   The breaking of a covalent chemical bond with the subsequent addition of a molecule of water.

hydrophilic   A polar compound that mixes readily with water.

hydrophobic   A nonpolar molecule that dissolves in fat and lipid solutions, but not in water.

hydroxyl group (-OH)   Chemical group consisting of oxygen and hydrogen that is a prominent part of alcohol.

image analysis   A computerized method for extracting information from digitized microscopic images of cells or cell organelles.

immunofluorescence   Detection of a specific cellular protein with the aid of a fluorescent dye that is coupled to an antibody.

immunoglobulin (Ig)   An antibody made by B cells as part of the adaptive immune response.

incontinence   Inability to control the flow of urine from the bladder (urinary incontinence) or the escape of stool from the rectum (fecal incontinence).

insertional mutagenesis   Damage suffered by a gene when a virus or a jumping gene inserts itself into a chromosome.

in situ hybridization   A method for studying gene expression, whereby a labeled cDNA or RNA probe hybridizes to a specific mRNA in intact cells or tissues. The procedure is usually carried out on tissue sections or smears of individual cells.

insulin   Polypeptide hormone secreted by β (beta) cells in the vertebrate pancreas. Production of this hormone is regulated directly by the amount of glucose that is in the blood.

interleukin   A small protein hormone, secreted by lymphocytes, to activate and coordinate the adaptive immune response.

interphase   The period between each cell division, which includes the G_1, S, and G_2 phases of the cell cycle.

intron   A section of a eukaryotic gene that is noncoding. It is transcribed but does not appear in the mature mRNA.

in vitro   Refers to cells growing in culture or a biochemical reaction occurring in a test tube (Latin for "in glass").

in vivo   A biochemical reaction, or a process, occurring in living cells or a living organism (Latin for "in life").

ion   An atom that has gained or lost electrons, thus acquiring a charge. Common examples are Na^+ and Ca^{++} ions.

ion channel   A transmembrane channel that allows ions to diffuse across the membrane down their electrochemical gradient.

ischemia   An inadequate supply of blood to a part of the body caused by degenerative vascular disease.

Jak-STAT signaling pathway   One of several cell signaling pathways that activates gene expression. The pathway is activated through cell surface receptors and cytoplasmic Janus kinases (Jaks) and signal transducers and activators of transcription (STATs).

karyotype   A pictorial catalogue of a cell's chromosomes, showing their number, size, shape, and overall banding pattern.

keratin   Proteins produced by specialized epithelial cells called keratinocytes. Keratin is found in hair, fingernails, and feathers.

kilometer   1,000 meters, which is equal to 0.621 miles.

kinesin   A motor protein that uses energy obtained from the hydrolysis of ATP to move along a microtubule.

kinetochore   A complex of proteins that forms around the centromere of mitotic or meiotic chromosomes, providing an attachment site for microtubules. The other end of each microtubule is attached to a chromosome.

Krebs cycle (citric acid cycle)   The central metabolic pathway in all eukaryotes and aerobic prokaryotes; discovered by the German chemist Hans Krebs in 1937. The cycle oxidizes acetyl groups derived from food molecules. The end products are CO_2, H_2O, and high-energy electrons, which pass via NADH and FADH2 to the respiratory chain. In eukaryotes, the Krebs cycle is located in the mitochondria.

labeling reaction   The addition of a radioactive atom or fluorescent dye to DNA or RNA for use as a probe in filter hybridization.

lagging strand   One of the two newly synthesized DNA strands at a replication fork. The lagging strand is synthesized discontinuously and therefore its completion lags behind the second, or leading, strand.

lambda bacteriophage   A viral parasite that infects bacteria; widely used as a DNA cloning vector.

leading strand   One of the two newly synthesized DNA strands at a replication fork. The leading strand is made by continuous synthesis in the 5' to 3' direction.

leucine zipper   A structural motif of DNA binding proteins, in which two identical proteins are joined together at regularly spaced leucine residues, much like a zipper, to form a dimer.

leukemia   Cancer of white blood cells.

lipid bilayer   Two closely aligned sheets of phospholipids that form the core structure of all cell membranes. The two layers are aligned such that the hydrophobic tails are interior, while the hydrophilic head groups are exterior on both surfaces.

liposome   An artificial lipid bilayer vesicle used in membrane studies and as an artificial gene therapy vector.

locus   A term from genetics that refers to the position of a gene along a chromosome. Different alleles of the same gene occupy the same locus.

long-term potentiation (LTP)   A physical remodeling of synaptic junctions that receive continuous stimulation.

lumen   A cavity completely surrounded by epithelial cells.

lymphocyte   A type of white blood cell that is involved in the adaptive immune response. There are two kinds of lymphocytes: T lymphocytes and B lymphocytes. T lymphocytes (T cells) mature in the thymus and attack invading microbes directly. B lymphocytes (B cells) mature in the bone marrow and make antibodies that are designed to immobilize or destroy specific microbes or antigens.

lysis   The rupture of the cell membrane followed by death of the cell.

lysosome Membrane-bounded organelle of eukaryotes that contains powerful digestive enzymes.

macromolecule A very large molecule that is built from smaller molecular subunits. Common examples are DNA, proteins, and polysaccharides.

magnetic resonance imaging (MRI) A procedure in which radio waves and a powerful magnet linked to a computer are used to create detailed pictures of areas inside the body. These pictures can show the difference between normal and diseased tissue. MRI makes better images of organs and soft tissue than other scanning techniques, such as CT or X-ray. MRI is especially useful for imaging the brain, spine, the soft tissue of joints, and the inside of bones. Also called nuclear magnetic resonance imaging.

major histocompatibility complex Vertebrate genes that code for a large family of cell-surface glycoproteins that bind foreign antigens and present them to T cells to induce an immune response.

malignant Refers to the functional status of a cancer cell that grows aggressively and is able to metastasize, or colonize, other areas of the body.

mammography The use of X-rays to create a picture of the breast.

MAP-kinase (mitogen-activated protein kinase) A protein kinase that is part of a cell proliferation–inducing signaling pathway.

M-cyclin A eukaryote enzyme that regulates mitosis.

meiosis A special form of cell division by which haploid gametes are produced. This is accomplished with two rounds of cell division, but only one round of DNA replication.

melanocyte A skin cell that produces the pigment melanin.

membrane The lipid bilayer and the associated glycocalyx that surround and enclose all cells.

membrane channel A protein complex that forms a pore or channel through the membrane for the free passage of ions and small molecules.

membrane potential A buildup of charged ions on one side of the cell membrane establishes an electrochemical gradient that is measured in millivolts (mV); an important characteristic of neurons as it provides the electrical current, when ion channels open, that enable these cells to communicate with one another.

mesoderm   An embryonic germ layer that gives rise to muscle, connective tissue, bones, and many internal organs.

messenger RNA (mRNA)   An RNA transcribed from a gene that is used as the gene template by the ribosomes and other components of the translation machinery to synthesize a protein.

metabolism   The sum total of the chemical processes that occur in living cells.

metaphase   The stage of mitosis at which the chromosomes are attached to the spindle but have not begun to move apart.

metaphase plate   Refers to the imaginary plane established by the chromosomes as they line up at right angles to the spindle poles.

metaplasia   A change in the pattern of cellular behavior that often precedes the development of cancer.

metastasis   Spread of cancer cells from the site of the original tumor to other parts of the body.

meter   Basic unit in the metric system; equal to 39.4 inches or 1.09 yards.

methyl group (-CH$_3$)   Hydrophobic chemical group derived from methane; occurs at the end of a fatty acid.

micrograph   Photograph taken through a light, or electron, microscope.

micrometer (μm or micron)   Equal to 10^{-6} meters.

microtubule   A fine cylindrical tube made of the protein tubulin, forming a major component of the eukaryote cytoskeleton.

millimeter (mm)   Equal to 10^{-3} meters.

mitochondrion (plural mitochondria)   Eukaryote organelle, formerly free living, that produces most of the cell's ATP.

mitogen   A hormone or signaling molecule that stimulates cells to grow and divide.

mitosis   Division of a eukaryotic nucleus; from the Greek *mitos,* meaning a thread, in reference to the threadlike appearance of interphase chromosomes.

mitotic chromosome   Highly condensed duplicated chromosomes held together by the centromere. Each member of the pair is referred to as a sister chromatid.

mitotic spindle   Array of microtubules, fanning out from the polar centrioles, and connecting to each of the chromosomes.

molecule   Two or more atoms linked together by covalent bonds.

monoclonal antibody   An antibody produced from a B cell–derived clonal line. Since all of the cells are clones of the original B cell, the antibodies produced are identical.

monocyte   A type of white blood cell that is involved in the immune response.

motif   An element of structure or pattern that may be a recurring domain in a variety of proteins.

M phase   The period of the cell cycle (mitosis or meiosis) when the chromosomes separate and migrate to the opposite poles of the spindle.

multipass transmembrane protein   A membrane protein that passes back and forth across the lipid bilayer.

multipotency   The property by which an undifferentiated animal cell can give rise to many of the body's cell types.

mutant   A genetic variation within a population.

mutation   A heritable change in the nucleotide sequence of a chromosome.

myelin sheath   Insulation applied to the axons of neurons. The sheath is produced by oligodendrocytes in the central nervous system and by Schwann cells in the peripheral nervous system.

myeloid cell   White blood cells other than lymphocytes.

myoblast   Muscle precursor cell; many myoblasts fuse into a syncytium, containing many nuclei, to form a single muscle cell.

myocyte   A muscle cell.

NAD (nicotine adenine dinucleotide)   Accepts a hydride ion (H^-), produced by the Krebs cycle, forming NADH, the main carrier of electrons for oxidative phosphorylation.

NADHdehydrogenase   Removes electrons from NADH and passes them down the electron transport chain.

nanometer (nm)   Equal to 10^{-9} meters or 10^{-3} microns.

National Institutes of Health (NIH)   A biomedical research center that is part of the U.S. Department of Health and Human Services. NIH consists of more than 25 research institutes, including the National Institute of Aging (NIA) and the National Cancer Institute (NCI). All of the institutes are funded by the federal government.

natural killer cell (NK cell)   A lymphocyte that kills virus-infected cells in the body; also kills foreign cells associated with a tissue or organ transplant.

neuromodulator   A chemical released by neurons at a synapse that modifies the behavior of the targeted neuron(s).

neuromuscular junction   A special form of synapse between a motor neuron and a skeletal muscle cell.

neuron   A cell specially adapted for communication that forms the nervous system of all animals.

neurotransmitter2   A chemical released by the synapse that activates the targeted neuron.

non−small cell lung cancer   A group of lung cancers that includes squamous cell carcinoma, adenocarcinoma, and large cell carcinoma. The small cells are endocrine cells.

northern blotting   A technique for the study of gene expression. Messenger RNA (mRNA) is fractionated on an agarose gel and then transferred to a piece of nylon filter paper (or membrane). A specific mRNA is detected by hybridization with a labeled DNA or RNA probe. The original blotting technique invented by E. M. Southern inspired the name. Also known as RNA blotting.

nuclear envelope   The double membrane (two lipid bilayers) enclosing the cell nucleus.

nuclear localization signal (NLS)   A short amino acid sequence located on proteins that are destined for the cell nucleus, after they are translated in the cytoplasm.

nuclei acid   DNA or RNA, a macromolecule consisting of a chain of nucleotides.

nucleolar organizer   Region of a chromosome containing a cluster of ribosomal RNA genes that gives rise to the nucleolus.

nucleolus   A structure in the nucleus where ribosomal RNA is transcribed and ribosomal subunits are assembled.

nucleoside   A purine or pyrimidine linked to a ribose or deoxyribose sugar.

nucleosome   A beadlike structure, consisting of histone proteins.

nucleotide   A nucleoside containing one or more phosphate groups linked to the 5' carbon of the ribose sugar. DNA and RNA are nucleotide polymers.

nucleus   Eukaryote cell organelle that contains the DNA genome on one or more chromosomes.

oligodendrocyte   A myelinating glia cell of the vertebrate central nervous system.

oligo labeling   A method for incorporating labeled nucleotides into a short piece of DNA or RNA. Also known as the random-primer labeling method.

oligomer   A short polymer, usually consisting of amino acids (oligopeptides), sugars (oligosaccharides), or nucleotides (oligonucleotides); taken from the Greek word *oligos*, meaning few or little.

oncogene   A mutant form of a normal cellular gene, known as a proto-oncogene, that can transform a cell to a cancerous phenotype.

oocyte   A female gamete or egg cell.

operator   A region of a prokaryote chromosome that controls the expression of adjacent genes.

operon   Two or more prokaryote genes that are transcribed into a single mRNA.

organelle   A membrane-bounded structure, occurring in eukaryote cells, that has a specialized function. Examples are the nucleus, Golgi complex, and endoplasmic reticulum.

osmosis   The movement of solvent across a semipermeable membrane that separates a solution with a high concentration of solutes from one with a low concentration of solutes. The membrane must be permeable to the solvent but not to the solutes. In the context of cellular osmosis, the solvent is always water, the solutes are ions and molecules, and the membrane is the cell membrane.

osteoblast   Cells that form bones.

ovulation   Rupture of a mature follicle with subsequent release of a mature oocyte from the ovary.

oxidative phosphorylation   Generation of high-energy electrons from food molecules that are used to power the synthesis of ATP from ADP and inorganic phosphate. The electrons are eventually transferred to oxygen, to complete the process; occurs in bacteria and mitochondria.

p53   A tumor suppressor gene that is mutated in about half of all human cancers. The normal function of the *p53* protein is to block passage through the cell cycle when DNA damage is detected.

parthenogenesis A natural form of animal cloning whereby an individual is produced without the formation of haploid gametes and the fertilization of an egg.

pathogen An organism that causes disease.

PCR (polymerase chain reaction) A method for amplifying specific regions of DNA by temperature cycling a reaction mixture containing the template, a heat-stable DNA polymerase, and replication primers.

peptide bond The chemical bond that links amino acids together to form a protein.

pH Measures the acidity of a solution as a negative logarithmic function (p) of H^+ concentration (H). Thus, a pH of 2.0 (10^{-2} molar H^+) is acidic, whereas a pH of 8.0 (10^{-8} molar H^+) is basic.

phagocyte A cell that engulfs other cells or debris by phagocytosis.

phagocytosis A process whereby cells engulf other cells or organic material by endocytosis. A common practice among protozoans and cells of the vertebrate immune system; from the Greek *phagein,* "to eat."

phenotype Physical characteristics of a cell or organism.

phosphokinase An enzyme that adds phosphate to proteins.

phospholipid The kind of lipid molecule used to construct cell membranes. Composed of a hydrophilic head-group, phosphate, glycerol, and two hydrophobic fatty acid tails.

phosphorylation A chemical reaction in which a phosphate is covalently bonded to another molecule.

photoreceptor A molecule or cell that responds to light.

photosynthesis A biochemical process in which plants, algae, and certain bacteria use energy obtained from sunlight to synthesize macromolecules from CO_2 and H_2O.

phylogeny The evolutionary history of a group of organisms, usually represented diagrammatically as a phylogenetic tree.

pinocytosis A form of endocytosis whereby fluid is brought into the cell from the environment.

pixel One element in a data array that represents an image or photograph.

placebo An inactive substance that looks the same and is administered in the same way as a drug in a clinical trial.

plasmid   A minichromosome, often carrying antibiotic-resistant genes, that occurs naturally among prokaryotes; used extensively as a DNA cloning vector.

platelet   A cell fragment derived from megakaryocytes and lacking a nucleus that is present in the bloodstream and is involved in blood coagulation.

ploidy   The total number of chromosomes (n) that a cell has. Ploidy is also measured as the amount of DNA (C) in a given cell, relative to a haploid nucleus of the same organism. Most organisms are diploid, having two sets of chromosomes, one from each parent, but there is great variation among plants and animals. The silk gland of the moth *Bombyx mori,* for example, has cells that are extremely polyploid, reaching values of 100,000C, flowers are often highly polyploid, and vertebrate hepatocytes may be 16C.

pluripotency   The property by which an undifferentiated animal cell can give rise to most of the body's cell types.

poikilotherm   An animal incapable of regulating its body temperature independent of the external environment. It is for this reason that such animals are restricted to warm tropical climates.

point mutation   A change in DNA, particularly in a region containing a gene, that alters a single nucleotide.

polarization   A term used to describe the reestablishment of a sodium ion gradient across the membrane of a neuron. Polarization followed by depolarization is the fundamental mechanism by which neurons communicate with one another.

polyacrylamide   A tough polymer gel that is used to fractionate DNA and protein samples.

polyploid   Possessing more than two sets of homologous chromosomes.

polyploidization   DNA replication in the absence of cell division; provides many copies of particular genes and thus occurs in cells that highly active metabolically (see ploidy).

portal system   A system of liver vessels that carries liver enzymes directly to the digestive tract.

post-mitotic   Refers to a cell that has lost the ability to divide.

probe   Usually a fragment of a cloned DNA molecule that is labeled with a radioisotope or fluorescent dye, and used to detect specific DNA or RNA molecules on southern or northern blots.

progenitor cell   A cell that has developed from a stem cell but can still give rise to a limited variety of cell types.

proliferation   A process whereby cells grow and divide.

promoter   A DNA sequence to which RNA polymerase binds to initiate gene transcription.

prophase   The first stage of mitosis; the chromosomes are duplicated and are beginning to condense but are attached to the spindle.

protein   A major constituent of cells and organisms. Proteins, made by linking amino acids together, are used for structural purposes and regulate many biochemical reactions in their alternative role as enzymes. Proteins range in size from just a few amino acids to more than 200.

protein glycosylation   The addition of sugar molecules to a protein.

proto-oncogene   A normal gene that can be converted to a cancer-causing gene (oncogene) by a point mutation or through inappropriate expression.

protozoa   Free-living, single-cell eukaryotes that feed on bacteria and other microorganisms. Common examples are *Paramecium* and *Amoeba*. Parasitic forms inhabit the digestive and urogenital tract of many animals, including humans.

P-site   The binding site on the ribosome for the growing protein (or peptide) chain.

purine   A nitrogen-containing compound that is found in RNA and DNA. Two examples are adenine and guanine.

pyrimidine   A nitrogen-containing compound found in RNA and DNA. Examples are cytosine, thymine, and uracil (RNA only).

radioactive isotope   An atom with an unstable nucleus that emits radiation as it decays.

randomized clinical trial   A study in which the participants are assigned by chance to separate groups that compare different treatments; neither the researchers nor the participants can choose which group. Using chance to assign people to groups means that the groups will be similar and that the treatments they receive can be compared objectively. At the time of the trial, it is not known which treatment is best.

random primer labeling   A method for incorporating labeled nucleotides into a short piece of DNA or RNA.

reagent A chemical solution designed for a specific biochemical or histochemical procedure.

recombinant DNA A DNA molecule that has been formed by joining two or more fragments from different sources.

refractive index A measure of the ability of a substance to bend a beam of light expressed in reference to air that has, by definition, a refractive index of 1.0.

regulatory sequence A DNA sequence to which proteins bind that regulate the assembly of the transcriptional machinery.

replication bubble Local dissociation of the DNA double helix in preparation for replication. Each bubble contains two replication forks.

replication fork The Y-shape region of a replicating chromosome; associated with replication bubbles.

replication origin (origin of replication, ORI) The location at which DNA replication begins.

respiratory chain (electron transport chain) A collection of iron- and copper-containing proteins, located in the inner mitochondrion membrane, that use the energy of electrons traveling down the chain to synthesize ATP.

restriction enzyme An enzyme that cuts DNA at specific sites.

restriction map The size and number of DNA fragments obtained after digesting with one or more restriction enzymes.

retrovirus A virus that converts its RNA genome to DNA once it has infected a cell.

reverse transcriptase An RNA-dependent DNA polymerase. This enzyme synthesizes DNA by using RNA as a template, the reverse of the usual flow of genetic information from DNA to RNA.

ribosomal RNA (rRNA) RNA that is part of the ribosome and serves both a structural and functional role, possibly by catalyzing some of the steps involved in protein synthesis.

ribosome A complex of protein and RNA that catalyzes the synthesis of proteins.

rough endoplasmic reticulum (rough ER) Endoplasmic reticulum that has ribosomes bound to its outer surface.

Saccharomyces Genus of budding yeast that are frequently used in the study of eukaryote cell biology.

sarcoma   Cancer of connective tissue.

Schwann cell   Glia cell that produces myelin in the peripheral nervous system.

screening   Checking for disease when there are no symptoms.

senescence   Physical and biochemical changes that occur in cells and organisms with age; from the Latin word *senex,* meaning "old man" or "old age."

signal transduction   A process by which a signal is relayed to the interior of a cell where it elicits a response at the cytoplasmic or nuclear level.

smooth muscle cell   Muscles lining the intestinal tract and arteries; lack the striations typical of cardiac and skeletal muscle, giving a smooth appearance when viewed under a microscope.

somatic cell   Any cell in a plant or animal except those that produce gametes (germ cells or germ cell precursors).

somatic cell nuclear transfer   Animal cloning technique whereby a somatic cell nucleus is transferred to an enucleated oocyte. Synonymous with cell nuclear transfer or replacement.

southern transfer   The transfer of DNA fragments from an agarose gel to a piece of nylon filter paper. Specific fragments are identified by hybridizing the filter to a labeled probe; invented by the Scottish scientist E. M. Southern, in 1975; also known as DNA blotting.

stem cell   Pluripotent progenitor cell found in embryos and various parts of the body that can differentiate into a wide variety of cell types.

steroid   A hydrophobic molecule with a characteristic four-ringed structure. Sex hormones, such as estrogen and testosterone, are steroids.

structural gene   A gene that codes for a protein or an RNA; distinguished from regions of the DNA that are involved in regulating gene expression but are noncoding.

synapse   A neural communication junction between an axon and a dendrite. Signal transmission occurs when neurotransmitters, released into the junction by the axon of one neuron, stimulate receptors on the dendrite of a second neuron.

syncytium   A large multinucleated cell. Skeletal muscle cells are syncytiums produced by the fusion of many myoblasts.

syngeneic transplants   A patient receives tissue or an organ from an identical twin.

tamoxifen   A drug that is used to treat breast cancer. Tamoxifen blocks the effects of the hormone estrogen in the body. It belongs to the family of drugs called antiestrogens.

T cell (T lymphocyte)   A white blood cell involved in activating and coordinating the immune response.

telomere   The end of a chromosome; replaced by the enzyme telomerase with each round of cell division to prevent shortening of the chromosomes.

telophase   The final stage of mitosis in which the chromosomes decondense and the nuclear envelope reforms.

template   A single strand of DNA or RNA whose sequence serves as a guide for the synthesis of a complementary, or daughter, strand.

therapeutic cloning   The cloning of a human embryo for the purpose of harvesting the inner cell mass (embryonic stem cells).

topoisomerase   An enzyme that makes reversible cuts in DNA to relieve strain or to undo knots.

totipotency   The property by which an undifferentiated animal cell can give rise to all of the body's cell types. The fertilized egg and blastomeres from an early embryo are the only cells possessing this ability.

transcription   The copying of a DNA sequence into RNA, catalyzed by RNA polymerase.

transcription factor   A general term referring to a wide assortment of proteins needed to initiate or regulate transcription.

transfection   Introduction of a foreign gene into a eukaryote or prokaryote cell.

transfer RNA (tRNA)   A collection of small RNA molecules that transfer an amino acid to a growing polypeptide chain on a ribosome. There is a separate tRNA for amino acid.

transgenic organism   A plant or animal that has been transfected with a foreign gene.

trans Golgi network   The membrane surfaces where glycoproteins and glycolipids exit the Golgi complex in transport vesicles.

translation   A ribosome-catalyzed process whereby the nucleotide sequence of a mRNA is used as a template to direct the synthesis of a protein.

transposable element (transposon)   A segment of DNA that can move from one region of a genome to another.

ultrasound (ultrasonography)   A procedure in which high-energy sound waves (ultrasound) are bounced off internal tissues or organs producing echoes that are used to form a picture of body tissues (a sonogram).

umbilical cord blood stem cells   Stem cells, produced by a human fetus and the placenta, that are found in the blood that passes from the placenta to the fetus.

vector   A virus or plasmid used to carry a DNA fragment into a bacterial cell (for cloning) or into a eukaryote to produce a transgenic organism.

vesicle   A membrane-bounded bubble found in eukaryote cells. Vesicles carry material from the ER to the Golgi and from the Golgi to the cell membrane.

virus   A particle containing an RNA or DNA genome surrounded by a protein coat. Viruses are cellular parasites that cause many diseases.

western blotting   The transfer of protein from a polyacrylamide gel to a piece of nylon filter paper. Specific proteins are detected with labeled antibodies. The name was inspired by the original blotting technique invented by the Scottish scientist E. M. Southern in 1975; also known as protein blotting.

xenogeneic transplants (xenograft)   A patient receives tissue or an organ from an animal of a different species.

yeast   Common term for unicellular eukaryotes that are used to brew beer and make bread. *Saccharomyces cerevisiae* (baker's yeast) are also widely used in studies on cell biology.

zygote   A diploid cell produced by the fusion of a sperm and egg.

 **Further Resources**

BOOKS

Alberts, Bruce, Dennis Bray, Karen Hopkins, and Alexander Johnson. *Essential Cell Biology.* 2d ed. New York: Garland, 2003. An introduction to cellular structure and function that is suitable for high school students.

Alberts, Bruce, Alexander Johnson, Julian Lewis, Martin Raff, Keith Roberts, and Peter Walter. *Molecular Biology of the Cell.* 5th ed. New York: Garland, 2008. Advanced coverage of cell biology that is suitable for senior high school students and undergraduates.

Boyles, Peter, and Bernard Levin, eds. *The World Cancer Report 2008.* Lyon, France: The International Agency for Research on Cancer, 2008. Available online. URL: http://www.iarc.fr/. Accessed January 1, 2009. The definitive source for cancer data from around the world.

Ganong, William. *Review of Medical Physiology.* 22d ed. Edition. New York: McGraw-Hill, 2005. A well-written overview of human physiology, beginning with basic properties of cells and tissues.

de Grey, Aubrey, and Michael Rae. *Ending Aging: The Rejuvenation Breakthroughs That Could Reverse Human Aging in Our Lifetime.* New York: St. Martin's Griffin, 2008. A comprehensive coverage of human aging aimed at the general public, high school students, and undergraduates.

Krause, W. J. *Krause's Essential Human Histology for Medical Students.* Boca Raton, Fla.: Universal Publishers, 2005. This book goes well with histology videos provided free on Google video.

Margulis, L., and Dorion Sagan. *Dazzle Gradually: Reflections on the Nature of Nature.* White River Junction, Vt.: Chelsea Green, 2007. Examines many interesting aspects of cellular evolution.

Margulis, L., and K. V. Schwartz. *Five Kingdoms: An Illustrated Guide to Phyla of Life on Earth.* New York: Freeman, 1998. Describes the evolution of bacteria, archaea, plants, animals, and fungi.

Panno, Joseph. *Aging: Modern Theories and Therapies.* Rev. ed. New York: Facts On File, 2010. Explains why and how people age and how stem cells may be used to reverse or modify the process.

———. *Animal Cloning: The Science of Nuclear Transfer.* Rev. ed. New York: Facts On File, 2010. Medical applications of cloning technology are discussed including therapeutic cloning.

———. *Cancer: The Role of Genes, Lifestyle, and Environment.* Rev. ed. New York: Facts On File, 2010. The basic nature of cancer written for the general public and young students.

———. *The Cell: Nature's First Life-form.* Rev. ed. New York: Facts On File, 2010. Everything you need to know about the cell without having to read a 1000-page textbook.

———. *Gene Therapy: Treatments and Cures for Genetic Diseases.* Rev. ed. New York: Facts On File, 2010. Discusses not only the great potential of this therapy, but also its dangers and its many failures.

JOURNALS AND MAGAZINES

Aiuti, Alessandro, et al. "Gene Therapy for Immunodeficiency Due to Adenosine Deaminase Deficiency." *New England Journal of Medicine* 360 (1/29/09): 447–458. This paper, from an Italian group, is a four-year follow-up on 10 children treated for ADA-SCID with stem cell gene therapy.

Aoi, Takashi, et al. "Generation of Pluripotent Stem Cells from Adult Mouse Liver and Stomach Cells." Sciencexpress (2/14/08): 1–4. Yamanaka's team characterizes the gene expression of iPS cells.

Brunet, Anne, and Thomas A. Rando. "Ageing: From Stem to Stern." *Nature* 449 (2007): 288–289. This article discusses the topic of immortality and suggests that important insights may be obtained by studying aging in stem cells.

Chung, Young, et al. "Reprogramming of Human Somatic Cells Using Human and Animal Oocytes" *Cloning and Stem Cells* 11 (2009): 1–11. This work, headed by Robert Lanza, shows that hybrid embryos fail to develop normally.

Church, George. "Genomes for All." *Scientific American* 294 (2006): 46–54. This article discusses fast and cheap DNA sequencers that could make it possible for everyone to have their genome sequenced, giving new meaning to personalized medicine.

Cibeli, Jose. "Is Therapeutic Cloning Dead?" *Science* 318 (12/21/07) 1,879–1,880. The author of this news article argues that since the creation of iPS cells, therapeutic cloning is no longer needed.

Collins, Francis, Michael Morgan, and Aristides Patrinos. "The Human Genome Project: Lessons from Large-Scale Biology." *Science* 300 (2003): 286–290. Provides an overview of the many organizational problems that had to be overcome in order to complete the project.

Costello, Joseph. "Tips for Priming Potency." *Nature* 454 (7/3/08): 45–46. A news article regarding the problems scientists encounter when producing iPS cells. This article serves as an introduction to a research article in the same issue (see Mikkelsen et al. below).

Dimos, John, et al. "Induced Pluripotent Stem Cells Generated from Patients with ALS Can Be Differentiated into Motor Neurons." Sciencexpress (7/31/08): 1–4. Kevin Eggan's team succeeded in producing an iPS-based therapy for a neurological disorder.

Gaspar, Robert, et al. "Successful Reconstitution of Immunity in ADA-SCID by Stem Cell Gene Therapy Following Cessation of PEG-ADA and Use of Mild Preconditioning." *Molecular Therapy* (10/4/06): 505–513. This British team used a combination of stem cells and gene therapy to treat ADA-SCID.

Guerra-Crespo, M., et al. "Transforming Growth Factor-alpha Induces Neurogenesis and Behavioral Improvement in a Chronic Stroke Model." Neuroscience. (2/24/09). A study conducted by James Fallon's group at the University of California, Irvine, in which a growth factor is used to stimulate stem cells to repair a rat's brain damaged by stroke.

Kaji, Keisuke, et al. "Virus-free Induction of Pluripotency and Subsequent Excision of Reprogramming Factors." *Nature* (3/1/09): 1–5. This paper describes a method for producing iPS cells without leaving the viral vectors inside the cells.

Kim, Jeong Beom, et al. "Pluripotent Stem Cells Induced from Adult Neural Stem Cells by Reprogramming with Two Factors." *Nature* 454 (7/31/08): 646–651.

Leong, Kevin, Bu-Er Wang, Leisa Johnson, and Wei-Quang Gao. "Generation of a Prostate from a Single Adult Stem Cell." *Nature* 456 (2008): 804–808. These authors identified markers specific for mouse prostate stem cells (PSCs). Using FACS, they isolated prostate PSCs and were able to grow a functional prostate from just one of these cells.

Mikkelsen, Tarjei, et al. "Dissecting Direct Reprogramming through Integrative Genomic Analysis." *Nature* 454 (7/3/08): 49–55. This paper characterizes some of the molecular events associated with the production of iPS cells.

Moreno-Manzano, V., et al. "Activated Spinal Cord Ependymal Stem Cells Rescue Neurological Function." *Stem Cells* 27 (2009): 733–743. A Spanish team, headed by Mirodrag Stojkovic, has produced an effective method to treat spinal cord injury with AS cells.

Orlic, Donald, et al. "Bone Marrow Cells Regenerate Infarcted Myocardium." *Nature* 410 (4/5/01): 701–705. Conducted at the New York Medical College, this is one of the earliest attempts to treat heart failure with stem cells.

Prassier, Robert, Linda van Laake, and Christine Mummery. "Stem Cell–based Therapy and Lessons from the Heart." *Nature* 414 (5/15/08): 322–329. Stem cell researchers have learned a lot by examining some of the failed attempts to treat cardiovascular disease with this therapy.

Ross, Joseph, et al. "Guest Authorship and Ghostwriting in Publications Related to Rofecoxib: A Case Study of Industry Documents from Rofecoxib Litigation." *Journal of the American Medical Association* 299 (4/16/08): 1,800–1,812. This report provides evidence in support of the accusation that Merk & Co. hired ghostwriters to produce papers that made their products look safe and effective.

Segers, Vincent, and Richard Lee. "Stem Cell Therapy for Cardiac Disease." *Nature* 451 (2/21/08): 937–942. This is a review article that covers most of the stem cell trials over the past eight years.

Soldner, Frank, et al. "Parkinson's Disease Patient–Derived Induced Pluripotent Stem Cells Free of Viral Reprogramming Factors." *Cell* 136 (3/6/09): 964–977. Rudolf Jaenisch's team, at the Whitehead Institute, Cambridge, Massachusetts, reprogram skin cells from a PD patient to produce virus-free iPS cells, which are then stimulated to differentiate into dopaminergic neurons.

Surani, Azim, and Anne McLaren. "A New Route to Rejuvenation." *Nature* 443 (9/21/06): 284–285. A news article that discusses the potential of iPS cells.

Takahashi, Kazutoshi, and Shinya Yamanaka. "Induction of Pluripotent Stem Cells from Mouse Embryonic and Adult Fibroblast Cultures by Defined Factors." *Cell* 126 (8/25/06): 663–676. The first report to describe the production of iPS cells.

Takahashi, Kazutoshi, et al. "Induction of Pluripotent Stem Cells from Adult Human Fibroblasts by Defined Factors." *Cell* 131 (11/30/07): 861–872. Yamanaka's team extend their work by producing iPS cells from human fibroblasts.

Wadman, Meredith. "Stem Cells Ready for Prime Time." *Nature* 457 (1/29/09): 516. The FDA approves the first clinical trial involving embryonic stem cells.

Wernig, Marius, et al. "In Vitro Reprogramming of Fibroblasts into a Pluripotent ES Cell–like State." *Nature* 448 (7/19/07): 318–325. Confirmation that iPS cells are similar to ES cells at the gene level.

———. "Neurons Derived from Reprogrammed Fibroblasts Functionally Integrate into the Fetal Brain and Improve Symptoms of Rats with Parkinson's Disease." *PNAS* 105 (4/7/08): 5,856–5,861. The first attempt to use iPS cells for a medical therapy.

Woltjen, Knut, et al. "PiggyBac Transposition Reprograms Fibroblasts to Induced Pluripotent Stem Cells." *Nature* (3/1/09): 1–5. A method for producing iPS cells that removes the viral vector after the cells have been reprogrammed.

Zhou, Qiao, et al. "In Vivo Reprogramming of Adult Pancreatic Exocrine Cells to β-cells." *Nature* 455 (10/2/08): 627–632. Sidestepping the use of stem cells, these researchers converted one type of pancreatic cell into another (the β-cells).

INTERNET ARTICLES

Associated Press. "Most Fertility Clinics Break the Rules." (2/20/09). Available online. URL: http://www.msnbc.msn.com/id/29305552/. Accessed March 9, 2009. This article discusses the practice of IVF clinics routinely producing more embryos than they need.

Baker, Monya. "Tumours Spark Stem Cell Review." *Nature News* (2/17/09). Available online. URL: http://www.nature.com/news/2009/090217/full/457941a.html. Accessed March 7, 2009.

An Israeli boy, treated with ES cells, developed tumors in his brain and along his spine that originated with the transplanted cells.

Clout, Laura. "Concern over Destroyed Embryos by IVF Clinics." *Telegraph* (12/31/07). Available online. URL: http://www. telegraph.co.uk/news/uknews/1574119/Concern-over-destroyed-embryos-by-IVF-c linics.html. Accessed March 9, 2009. Discusses the ethics of destroying extra embryos produced by IVF clinics.

Harris, Gardiner. "Some Stem Cell Research Limits Lifted." *New York Times* (4/17/09). Available online. URL: http://www. nytimes.com/2009/04/18/health/18stem.html?_r=1&ref=health. Accessed April 18, 2009. This article discusses Obama's executive order lifting some of the restrictions on ES cell research.

Hill, Amelia. "Private IVF Clinics Are 'Exploiting Women'" *Guardian* (7/15/07). Available online. URL: http://www.guardian. co.uk/society/2007/jul/15/health.medicineandhealth. Accessed March 9, 2009. IVF clinics tend to transplant too many embryos in order to ensure success, but the multiple births can jeopardize the health and safety of both the mother and the infants.

Morales, Lymari. "Majority of Americans Likely Support Stem Cell Decision." Gallup (3/9/09). Available online. URL: http://www. gallup.com/poll/116485/Majority-Americans-Likely-Support-Stem-Cell-Decision.aspx. Accessed March 10, 2009. The Gallup poll showed that while a slight majority support Obama's executive order, they also want some restrictions placed on the work.

National Institutes of Health. "Regenerative Medicine 2006." Available online. URL: http://stemcells.nih.gov/info/scireport/ 2006report.htm. Accessed January 1, 2009. A 62-page pamphlet, written for the general public, describing stem cell research and therapies that may be possible in the future.

Roan, Shari. "Cord-blood Banking: Worth It or Not?" *Los Angeles Times* (3/2/09). Available online. URL: http://www.latimes.com/ news/science/la-he-cordblooddonate2-2009mar02,0,2439927.

story. Accessed March 9, 2009. The article concludes that it is worth it, but that the blood should be stored at public banks.

Saul, Stephanie. "Birth of Octuplets Puts Focus on Fertility Clinics." *New York Times* (2/11/09). Available online. URL: http://www.nytimes.com/2009/02/12/health/12ivf.html?_r=1. Accessed March 11, 2009. This article discusses the practice at IVF clinics of producing too many embryos and then transplanting an excessive number to ensure success.

Whitehouse.gov. "The Briefing Room." Available online. URL: http://www.whitehouse.gov/the_press_office/Removing-Barriers-to-Responsible-Scientific-Research-Involving-Human-Stem-Cells/ Accessed April 10, 2009. The text of Obama's executive order that lifted some restrictions on embryonic stem cell research.

Yoshiro, Kimi, and Jessica Garrison. "Stricter Rules on Fertility Industry Debated." *Los Angeles Times* (3/6/09). Available online. URL: http://www.latimes.com/news/science/la-na-octuplets6-2009mar06,0,6503134.story. Accessed March 9, 2009. The stricter rules would limit the number of embryos that could be transplanted to two or three.

WEB SITES

Centers for Disease Control and Prevention (CDC). Available online. URL: http://www.cdc.gov/art/. Accessed March 10, 2009. The CDC is one of the major components of the Department of Health and Human Services. Its site contains a great deal of information about diseases that are common in North America.

Gallup. Available online. URL: http://www.gallup.com/home.aspx. Accessed November 7, 2009. Gallup is a news and polling organization that analyzes public opinion concerning a wide range of topics.

Genetic Science Learning Center at the Eccles Institute of Human Genetics, University of Utah. An excellent resource for beginning students. Available online. URL: http://learn.genetics.utah. edu/. Accessed April 27, 2009. This site contains information and illustrations covering basic cell biology, animal cloning, gene therapy, and stem cells.

Geron Corporation. Available online. URL: http://www.geron. com/grnopc1clearance/. Accessed March 10, 2009. The corporation has posted documents describing its clinical trial to treat spinal cord injury with ES cell-derived oligodendrocytes.

Google Video. Available online. URL: http://video.google.com/ videosearch?q=histology+tissue&emb=0&aq=3oq=histology#. Accessed April 27, 2009. This site contains many videos covering human histology and cell biology.

Guttmacher Institute. Available online. URL: http://www/guttmacher.org/. Accessed November 7, 2009. This institute analyzes all aspects of sexual and reproductive health worldwide.

Human Fertilization and Embryology Authority. Available online. URL: http://www.hfea.gov.uk/. Accessed March 9, 2009. The British government agency that regulates IVF clinics and all research involving human embryos.

Institute of Molecular Biotechnology. Jena, Germany. Available online. URL: http://www.imb-jena.de/IMAGE.html. Accessed April 27, 2009. This site has an image library of biological macromolecules.

National Center for Biotechnology Information (NCBI). Available online. URL: http://www.ncbi.nlm.nih.gov. Accessed April 23, 2009. This site, established by the National Institutes of Health, is an excellent resource for anyone interested in biology. The NCBI provides access to GenBank (DNA sequences), literature databases (Medline and others), molecular databases, and topics dealing with genomic biology. With the literature database, for example, anyone can access Medline's 11 million biomedical journal citations to research biomedical questions. Many of these links provide free access to full-length research papers.

National Health Museum Resource Center. Washington, D.C. Available online. URL: www.accessexcellence.org/RC/. Accessed April 23, 2009. Covers many areas of biological research, supplemented with extensive graphics and animations.

National Human Genome Research Institute United States. Available online. URL: http://www.genome.gov/. Accessed April 23, 2009. The institute supports genetic and genomic research, including the ethical, legal, and social implications of genetics research.

National Institutes of Health. Available online. URL: http://www.nih.gov. Accessed April 23, 2009. The NIH posts information on their Web site that covers a broad range of topics including general health information, stem cell biology, aging, cancer research, and much more.

Nature Publishing Group. Available online. URL: http://www.nature.com/nature/supplements/collections/humangenome/commentaries/. Accessed April 23, 2009. The journal *Nature* has provided a comprehensive guide to the human genome. This site provides links to the definitive historical record for the sequences and analyses of human chromosomes. All papers, which are free for downloading, are based on the final draft produced by the Human Genome Project.

Tufts Center for the Study of Drug Development. Available online. URL: http://csdd.tufts.edu/Default.asp. Accessed April 23, 2009. This site publishes bimonthly reports discussing the many issues associated with the development of new drugs and therapies.

U.S. Food and Drug Administration. Available online. URL: http://www.fda.gov. Accessed April 27, 2009. Provides extensive coverage of general health issues and regulations.

Index

Note: *Italic* page numbers indicate illustrations.

A

Aastrom Biosciences 119–120
abortion
    and British regulations 147, 151
    and European regulations 144–146,
        147, 154
    as source of embryos 41–42, 140–141
    and U.S. regulations 146–147, 156
Abortion Act (United Kingdom) 151
ACT. *See* Advanced Cell Technology
actin *167*
activation, cellular 185
activin-A 53, 54
ADA deficiency. *See* adenosine deaminase
    deficiency
ADA gene 99
adenine 172, 176, 178
adenosine deaminase (ADA) deficiency
    89, 99, 197
adenosine triphosphate (ATP) 169
    recycling of *98*
    synthesis of 179–180
adenovirus, as vector 195, *196*
adult stem (AS) cells 42–48, 160–162
    blood cell production from 54–55
    commercialization of 119–120
    definition of 42
    determining source of 42–44
    differentiation of 44–45

    directed 47, 51–60
    first characterized 42
    future prospects for 60–61
    identification of 42–44
    immune response to 46–47, 55, 128
    markers for 48–51
    medical applications of
        in Alzheimer's disease 107, 163
        in cardiovascular disease 90–93
        in diabetes 94
        in organ production 136
        in spinal cord injury 111–113
    plasticity of xv, 27, 44–45, 55–56
    research focus on 28–29
    tissues and organs known to have 43
Advanced Cell Technology (ACT) 41, 122,
    140, 141–142
advocates
    embryo and fetal 146–147
    patient 139–140
A gene, of *Lac* operon 79, *80*
AIDS dementia *109*
Alan Guttmacher Institute 146
algae, blue-green 7
allogeneic therapy 89, 92, 100, 126, 136
allografts 126, 136
α cells, of pancreas 93
ALS. *See* amyotrophic lateral sclerosis
Alzheimer, Alois 105
Alzheimer's disease 104–107
    cerebral atrophy in *109*
    genetics of 105
    progression of 104–105, *108*